Geometric Formulas

Rectangle
    Area: $A = \ell w$
    Perimeter: $P = 2\ell + 2w$

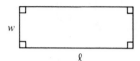

Square
    Area: $A = s^2$
    Perimeter: $P = 4s$

Parallelogram
    Area: $A = bh$

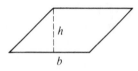

Trapezoid
    Area: $A = \frac{1}{2}h(a + b)$

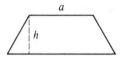

Triangle
    Area: $A = \frac{1}{2}bh$

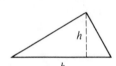 or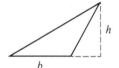

    Area: $A = \sqrt{s(s - a)(s - b)(s - c)}$,
    where $s = \frac{1}{2}(a + b + c)$
    Angle sum: $A + B + C = 180°$

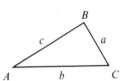

Right Triangle
    Pythagorean theorem: $a^2 + b^2 = c^2$

Circle
    Area: $A = \pi r^2$
    Circumference: $C = \pi d = 2\pi r$

Student Study Guide and Solutions Manual

Fundamentals of Algebra and Trigonometry

Dennis T. Christy
Nassau Community College

Deborah R. Levine
Nassau Community College

wcb
Wm. C. Brown Publishers
Dubuque, Iowa

Cover credit © 1988 Lehn & Associates, Inc.

ISBN 0–697–05626–0

Printed in the United States of America by Wm. C. Brown Publishers
2460 Kerper Boulevard, Dubuque, IA 52001

10 9 8 7 6 5 4 3 2 1

Contents

To the Student

This *Student Study Guide and Solutions Manual* is intended as a supplement to the text *Fundamentals of Algebra and Trigonometry* by Dennis T. Christy. Its purpose is twofold: first, to help you focus on the key objectives, terms, rules, and formulas given in the text and second, to encourage a learn-by-doing approach by providing detailed solutions to selected even-numbered exercises and additional margin exercises and sample test questions to check mastery of the basic concepts. To accomplish these objectives, the *Student Study Guide and Solutions Manual* covers the following key aspects for each section in the text.

1. Specific objectives for the section.
2. A list of important terms.
3. A summary of the key rules and formulas.
4. Additional comments (if needed) to point out subtleties in some of the rules or common mistakes to avoid.
5. Detailed solutions to selected even-numbered exercises.
6. Margin exercises matched to the solved problems.

The detailed solutions contain at least one problem from each exercise group, and an effort was made to select exercises that illustrate solutions different from the example problems in the text as well as solutions to some of the more challenging problems. By doing the matched margin exercises, you can get additional practice and check your progress.

The *Student Study Guide and Solutions Manual* also contains both a text overview that is composed from the chapter overviews in the text and sample test questions (with answers) for each chapter. The questions are presented in a variety of formats (such as multiple choice and completing the statement) and concentrate on the basic skills of evaluating expressions, graphing, solving equations and inequalities, applying algebraic and function laws, and setting up and solving applied problems. To prepare for your actual exams, it will always be necessary for you to combine proficiency in these basic areas with an understanding of the specific concepts and problems stressed in your classes.

A scientific calculator is frequently used in both the *Student Study Guide and Solutions Manual* and the text to evaluate a wide variety of expressions, and since it is inexpensive, easy to use, and a significant learning aid, we recommend that you obtain one. Calculator illustrations show primarily the keystrokes required on a Texas Instruments TI-30 SLR+, but for any model, you should read the owner's manual to familiarize yourself with its specific keys and limitations.

We hope this book becomes a valuable aid for you and helps you deepen and extend your understanding of mathematics.

Dennis T. Christy
Deborah R. Levine

Chapter 1
Fundamentals of Algebra

1.1 Real numbers

Objectives checklist

Can you:

a. Identify an integer, rational number, irrational number, and real number?

b. Express rational numbers as repeating decimals, and vice versa?

c. Identify the properties of real numbers?

d. Use the proper symbol ($<$, $>$, $=$) between a pair of numbers to indicate their correct order?

e. Find the absolute value of a real number?

f. Add, subtract, multiply, and divide real numbers?

Key terms

Real number
Rational number
Irrational number
Negative (or additive inverse) of a
Real number line
Absolute value
Factor
Power
Exponent

Integer
Positive integer
Subset
If and only if
Less than
Greater than
Less than or equal to
Greater than or equal to

Key rules and formulas

•   The relationship among the various sets of numbers is shown below.

$$\text{Real numbers} \begin{cases} \text{Rational numbers} \begin{cases} \text{Integers} \end{cases} \\ \text{Irrational numbers} \end{cases}$$

•   *Closure property of addition:* $a + b$ is a unique real number.

•   *Closure property of multiplication:* ab is a unique real number.

•   *Associative property of addition:* $(a + b) + c = a + (b + c)$. We obtain the same result if we change the grouping of real numbers in an addition problem.

•   *Associative property of multiplication:* $(a \cdot b) \cdot c = a \cdot (b \cdot c)$. We obtain the same result if we change the grouping of real numbers in a multiplication problem.

•   *Commutative property of addition:* $a + b = b + a$. The order in which we write real numbers in an addition problem does not affect their sum.

•   *Commutative property of multiplication:* $a \cdot b = b \cdot a$. The order in which we write real numbers in a multiplication problem does not affect their product.

•   *Inverse property of addition:* For every real number a, there is a unique real number, denoted by $-a$, such that $a + (-a) = (-a) + a = 0$.

Chapter 1

1.1

A. Express $\frac{7}{6}$ as a repeating decimal.

B. Classify the number $\sqrt{2}$ using the following categories: real number, rational number, irrational number, integer, positive integer, or none of these.

C. Classify $0.\overline{3}$ using the categories from Exercise B.

D. Name the property illustrated by $(1 + 2) + 3 = (2 + 1) + 3$.

E. Name the property illustrated by $(ax)y = a(xy)$.

- *Inverse property of multiplication:* For every real number a (except zero), there is a unique real number, denoted by $1/a$, such that $a \cdot (1/a) = (1/a) \cdot a = 1$.

- *Distributive properties:* $a(b + c) = ab + ac$ and $(a + b)c = ac + bc$

- *Trichotomy property:* If a and b are real numbers, then either $a > b$, $a < b$, or $a = b$.

- *Absolute value:* The absolute value of the number a is denoted by $|a|$ and is interpreted geometrically as the distance on the number line from a to 0. By definition,

$$|a| = \begin{cases} a \text{ if } a \geq 0 \\ -a \text{ if } a < 0. \end{cases}$$

- *Addition:*
  1. When adding two real numbers of the same sign, add their absolute values and use the sign they have in common.
  2. When adding two real numbers with different signs, subtract the smaller absolute value from the larger and use the sign of the number with the larger absolute value.

- *Subtraction:* If a and b are any real numbers, then $a - b = a + (-b)$.

- *Multiplication:*
  1. If the signs of two real numbers are different, the sign of the product is negative.
  2. If the signs of two real numbers are the same, the sign of the product is positive.

- *Division:* If a and b are any real numbers ($b \neq 0$), then $a \div b = a(1/b)$.

- Division by 0 is undefined.

- *Order of operations:* Do the operations in the following order:
  1. Perform all operations within parentheses.
  2. Find powers or roots.
  3. Perform all multiplications and divisions.
  4. Perform all additions and subtractions.

Detailed solutions to selected exercises

Exercise 6   Express $\frac{26}{11}$ as a repeating decimal.

Solution   Since $26 \div 11 = 2.3636\ldots$, we can express $\frac{26}{11}$ as $2.\overline{36}$. If you use a calculator for the division, you might see 2.3636364 and think the number is not a repeating decimal. However, the calculator is merely rounding off the last digit.
Now do **A.**

Exercise 20   Classify the number $\frac{0}{3}$ using the following categories: real number, rational number, irrational number, integer, positive integer, or none of these.

Solution   Since $\frac{0}{3} = 0$, the number $\frac{0}{3}$ is an integer, a rational number, and a real number. Note in the diagram of number relationships that any integer is automatically a rational number and a real number.
Now do **B.**

Exercise 30   Classify 0.101001 . . . using the categories from Exercise 20.

Solution   The number 0.101001 . . . is a nonrepeating decimal. Thus, it is an irrational number and a real number.
Now do **C**.

Exercise 40   Name the property of real numbers illustrated by the statement $(5 + 3) + (2 + 1) = [(5 + 3) + 2] + 1$.

Solution   On the left side of the equality 2 is grouped with 1, while on the right side 2 is grouped with $(5 + 3)$. Since the grouping is changed, this statement illustrates the associative property of addition.
Now do **D**.

Exercise 48   Name the property illustrated by $ax + ay = a(x + y)$.

Solution   The statement illustrates the distributive property. You may find it easier to spot this property if you rewrite the statement as
$a(x + y) = ax + ay$.
Now do **E**.

Exercise 60   Use the proper symbol $(<, >, =)$ in the statement
0.0001 _____ 0.00001.

Solution   On the number line the point representing 0.0001 is to the right of the point representing 0.00001. Therefore, 0.0001 > 0.00001.
Now do **F**.

Exercise 82   Evaluate $(-3.3) + (-0.67)$.

Solution $(-3.3) + (-0.67) = -3.97$ since $|-3.3| = 3.3, |-0.67| = 0.67$, $3.3 + 0.67 = 3.97$, and the sign that -3.3 and -0.67 have in common is negative.
Now do **G**.

Exercise 100   Evaluate $(-4 + 2)(3 - 8)$.

Solution $(-4 + 2)(3 - 8) = (-2)(-5) = 10$
Now do **H**.

Exercise 112   Evaluate $7 - 3[2(13 - 5) - (5 - 13)]$.

Solution $7 - 3[2(13 - 5) - (5 - 13)] = 7 - 3[2(8) - (-8)]$
$$= 7 - 3[24]$$
$$= 7 - 72$$
$$= -65$$
Now do **I**.

Exercise 122   True or False: All integers are irrational numbers.

Solution   False. The number 1 is an integer but not an irrational number. In fact, no integer is an irrational number.
Now do **J**.

1.2 Algebraic expressions and integer exponents

Objectives checklist

Can you:

a. Add and subtract algebraic expressions?

b. Remove parentheses or brackets in an algebraic expression?

F. Use the proper symbol $(>, <, =)$ in the statement -0.001 _____ -0.00001.

G. Evaluate $(-4.6) + (-0.23)$.

H. Evaluate $(-2 + 3)(5 - 1)$.

I. Evaluate
$3 - 4[2(4 - 5) - (3 + 7)]$.

J. True or false: All integers are real numbers.

c. State the number of terms in an algebraic expression?

d. Evaluate algebraic expressions, including expressions containing integer exponents?

e. Evaluate expressions using a scientific calculator?

f. Rewrite an expression with negative or zero exponents so it contains only positive exponents?

g. Use the laws of exponents to simplify certain products and quotients and to raise a power to a power?

Key terms

Variable	Subscript
Constant	Algebraic operating system (AOS)
Absolute constant	Terms
Arbitrary constant	Coefficient
Algebraic expression	Similar terms

Key rules and formulas

- According to the distributive property, we combine similar terms by combining their coefficients.

- Parentheses are removed by applying the distributive property. In the case of a grouping that is preceded by a minus sign, the factor is -1, so the sign of each term inside the parentheses must be changed.

- *Laws of exponents:* If m and n denote integers,

  1. $a^m \cdot a^n = a^{m+n}$
  2. $(a^m)^n = a^{mn}$
  3. a) $\dfrac{a^m}{a^n} = a^{m-n} \qquad (a \neq 0)$

     b) To obtain positive exponents directly, use

     $$\frac{a^m}{a^n} = \begin{cases} a^{m-n} & \text{if } m > n \\ 1/a^{n-m} & \text{if } n > m \end{cases} \qquad (a \neq 0).$$

- *Zero exponent:* $a^0 = 1 \qquad (a \neq 0)$

- *Negative exponent:* $a^{-n} = \left(\dfrac{1}{a}\right)^n = \dfrac{1}{a^n} \qquad (a \neq 0)$

- Any factor of the numerator may be made a factor of the denominator (and vice versa) by changing the sign of the exponent. This principle applies only to factors.

Additional comments

- In a fraction a negative sign may be placed in the following three positions:

  $$-\frac{a}{b} = \frac{-a}{b} = \frac{a}{-b}$$

- $-x^2$ means you must first square x and then take the negative of your result.

• *Calculator computation:* Consult the owner's manual for your calculator and the guidelines given in the text.

Detailed solutions to selected exercises

Exercise 12 If $x = -2$, $y = 3$, and $z = -4$, evaluate $(x - 2)(3y - 4z)$.

Solution
$$(x - 2)(3y - 4z) = (-2 - 2)(3(3) - 4(-4))$$
$$= (-4)(25)$$
$$= -100$$

Now do **A.**

Exercise 24 If $x_1 = 2$, $y_1 = -3$, $x_2 = 2$, and $y_2 = -6$, evaluate
$$m = \frac{y_2 - y_1}{x_2 - x_1}.$$

Solution $m = \dfrac{y_2 - y_1}{x_2 - x_1} = \dfrac{-6 - (-3)}{2 - (2)} = \dfrac{-6 + 3}{2 - 2} = \dfrac{-3}{0}$. Since division by 0 is undefined, m is undefined.
Now do **B.**

Exercise 42 Simplify $(3c^2d + 2cd - 5d^3) - (9d^3 - 6c^2d - 2cd)$.

Solution
$$3c^2d + 2cd - 5d^3$$
$$\text{add} \quad 6c^2d + 2cd - 9d^3 \qquad \text{The sign of each term is changed.}$$
$$9c^2d + 4cd - 14d^3$$
The answer $9c^2d + 4cd - 14d^3$ has three terms.
Now do **C.**

Exercise 48 Simplify $10 - 4[3x - (1 - x)]$.

Solution
$$10 - 4[3x - (1 - x)] = 10 - 4[3x - 1 + x]$$
$$= 10 - 4[4x - 1]$$
$$= 10 - 16x + 4$$
$$= 14 - 16x$$
The answer $14 - 16x$ has two terms. Note that $14 - 16x \neq -2x$.
Now do **D.**

Exercise 68 Evaluate $\dfrac{5^{-2}}{5^{-4}}$. Check your result with a calculator.

Solution By $\dfrac{a^m}{a^n} = a^{m-n}$, we have $\dfrac{5^{-2}}{5^{-4}} = 5^{-2 - (-4)} = 5^2 = 25$. An alternative method is to first rewrite the expression with only positive exponents so $\dfrac{5^{-2}}{5^{-4}} = \dfrac{5^4}{5^2} = 5^{4-2} = 5^2 = 25$.
Calculator check: $5\ \boxed{y^x}\ 2\ \boxed{+/-}\ \boxed{\div}\ 5\ \boxed{y^x}\ 4\ \boxed{+/-}\ \boxed{=}\ 25$.
Now do **E.**

Exercise 82 Simplify $(2x^{-3})(-3x^{-2})$. Write the result using only positive exponents.

Solution Here are two ways to solve the problem.

1. $(2x^{-3})(-3x^{-2}) = 2(-3)x^{-3} \cdot x^{-2} = -6x^{-3 + (-2)} = -6x^{-5} = -6/x^5$
2. $(2x^{-3})(-3x^{-2}) = (2/x^3)(-3/x^2) = 2 \cdot (-3)/(x^3 \cdot x^2) = -6/x^5$

1.2

A. If $x = -2$, $y = 3$, and $z = -4$, evaluate $(2 - x)(4z - 3y)$.

B. If $x_1 = -2$, $y_1 = 3$, $x_2 = 1$, and $y_2 = -5$, evaluate $m = \dfrac{y_2 - y_1}{x_2 - x_1}$.

C. Simplify $(2x^2y - 3xy + 5y^3) - (3y^3 - 5x^2y + xy)$.

D. Simplify $3 - 2[4x - (3 - x)]$.

E. Evaluate $\dfrac{6^{-3}}{6^{-7}}$.

6

F. Simplify $(-4x^{-2})(-3x^{-3})$ and write the result using positive exponents only.

Now do **F.**

Exercise 102   Simplify $5^{2x}/5^x$.

Solution   By $a^m/a^n = a^{m-n}$, we have $5^{2x}/5^x = 5^{2x-x} = 5^x$.
Now do **G.**

Exercise 114   Simplify $(2x^a)^c(3x^c)^b$.

Solution   First, remove parentheses. We raise a power to a power by multiplying exponents so $(2x^a)^c \cdot (3x^c)^b = 2^c \cdot x^{ac} \cdot 3^b \cdot x^{cb}$. Now add exponents from the factors with base x to obtain the answer $2^c \cdot 3^b \cdot x^{ac+cb}$.
Now do **H.**

G. Simplify $\dfrac{2^{5x}}{2^2}$.

Exercise 130   Does the keystroke sequence $2\;\boxed{y^x}\;3\;\boxed{\times}\;5\;\boxed{=}$ produce the correct answer to $2^{3(5)}$?

Solution   No. $2\;\boxed{y^x}\;3\;\boxed{\times}\;5\;\boxed{=}$ answers $2^3 \cdot 5$ because powers are computed before multiplication. To compute $2^{3(5)}$, use $2\;\boxed{y^x}\;\boxed{(}\;3\;\boxed{\times}\;5\;\boxed{)}\;\boxed{=}$.
Now do **I.**

Exercise 136   See text for question.

H. Simplify $4(y^m)^{-n} \cdot 2(y^m)^{-n}$.

Solution   Although you might expect the difference between the rope lengths to be very large, they only differ by about 6 ft. Since the equator of the earth may be approximated by a circle with diameter d, the circumference formula tells us that the length of the first rope is πd, while the length of the second rope is $\pi(d+2)$. Then

$$\text{difference} = \pi(d+2) - \pi d = \pi d + 2\pi - \pi d = 2\pi \text{ ft} \approx 6 \text{ ft.}$$

Now do **J.**

I. Does the keystroke sequence $4\;\boxed{y^x}\;2\;\boxed{\times}\;3\;\boxed{=}$ produce the correct answer to $(4^2)(3)$?

1.3 Products of algebraic expressions and factoring

Objectives checklist

Can you:

a. Use the distributive property to multiply an expression with one term by an expression with more than one term?

b. Use the distributive property twice to multiply an expression with two terms by an expression with at least two terms?

c. Use the FOIL method to multiply mentally two expressions that each contain two terms?

d. Factor an expression that contains a factor common to each term?

e. Factor an expression that is the difference of squares?

f. Factor an expression that is a trinomial?

g. Factor an expression that is the sum or difference of cubes?

h. Factor an expression by grouping together terms and using the methods specified above?

i. Determine if an expression of the form $ax^2 + bx + c$ can be factored?

J. A rope is wrapped tightly around the equator of the earth. A second rope is suspended in the air 2 ft directly above the first all the way around. To the nearest foot, what is the difference between the rope lengths?

Key terms

Factor completely
Greatest common factor
Polynomial
Monomial
Binomial

Trinomial
Degree of a monomial
Degree of a polynomial
Discriminant
Irreducible polynomial

Key rules and formulas

- If both factors in the multiplication contain more than one term, then we multiply each term of the first factor by each term of the second factor and then combine similar terms.

- The degree of a monomial is the sum of the exponents on the variables in the term.

- The degree of a polynomial is the same as the degree of its highest monomial term.

- *Factoring and product models:*

  1. Common factor: $ax + ay = a(x + y)$
  2. Difference of squares: $x^2 - y^2 = (x + y)(x - y)$
  3. Trinomial (leading coefficient 1):
     $x^2 + (a + b)x + ab = (x + a)(x + b)$
  4. Perfect square trinomial: $x^2 + 2xy + y^2 = (x + y)^2$
  5. General trinomial: $(ac)x^2 + (ad + bc)x + bd = (ax + b)(cx + d)$
  6. Sum of cubes: $a^3 + b^3 = (a + b)(a^2 - ab + b^2)$
  7. Difference of cubes: $a^3 - b^3 = (a - b)(a^2 + ab + b^2)$

- *Factoring test for trinomials:* A trinomial of the form $ax^2 + bx + c$, where a, b, and c are integers, is factorable into binomial factors with integer coefficients if and only if $b^2 - 4ac$ is a perfect square.

Additional comments

- The *First* + *Outer* + *Inner* + *Last*, or FOIL, method can be used when both factors contain two terms. This method is illustrated below.

$$(x^3 + y^3)(x^3 + y^3)$$

$$F + O + I + L$$
$$x^6 + x^3y^3 + x^3y^3 + y^6$$
$$= x^6 + 2x^3y^3 + y^6$$

  Note that it is often possible to combine the outer term and the inner term to form the middle term in the product.

- Since $a \div b = a \cdot (1/b)$, the distributive property can also be used to simplify problems involving division by one term.

- Unless otherwise specified, we only factor expressions into factors with integer coefficients.

- An expression is factored completely when none of the factors can be factored again.

1.3

A. Multiply $(6x - y)(2y - 3x)$.

B. Factor completely
$4a^7b^3 - 16a^5b^9 + 24ab^5$.

C. Factor completely
$144x^6 - 121a^4$.

D. Factor completely $x^2 + 2x - 15$.

Detailed solutions to selected exercises

Exercise 18   Multiply $(5y - 4x)(3y + x)$.

Solution   We multiply each term of the first factor by each term of the second factor in any of the following ways:

Method 1: $(5y - 4x)(3y + x) = (5y - 4x)(3y) + (5y - 4x)(x)$
$$= 15y^2 - 12xy + 5xy - 4x^2$$
$$= 15y^2 - 7xy - 4x^2$$

Method 2:
$$5y - 4x$$
$$3y + x$$
$$15y^2 - 12xy \qquad \text{This line equals } 3y(5y - 4x).$$
add $\quad \dfrac{+ \quad 5xy - 4x^2}{15y^2 - \quad 7xy - 4x^2} \qquad \text{This line equals } x(5y - 4x).$

Foil Method: $\qquad\qquad\quad$ F $\; + \;$ O $\; + \;$ I $\; + \;$ L
$$(5y - 4x)(3y + x) = 15y^2 + 5xy - 12xy - 4x^2 = 15y^2 - 7xy - 4x^2$$

Now do **A**.

Exercise 46   Factor completely $3a^{10}x^6 - 9a^7x^7 + 6a^9x^4$.

Solution   Greatest common factor: $3a^7x^4$. If $3a^7x^4$ is one factor, the other factor is

$$(3a^{10}x^6 - 9a^7x^7 + 6a^9x^4) \div 3a^7x^4 = a^3x^2 - 3x^3 + 2a^2.$$

Therefore, $3a^{10}x^6 - 9a^7x^7 + 6a^9x^4 = 3a^7x^4(a^3x^2 - 3x^3 + 2a^2)$.
Now do **B**.

Exercise 52   Factor completely $25x^{12} - 36a^{10}$.

Solution   The expression is the difference of squares with $25x^{12} = 5x^6 \cdot 5x^6$ and $36a^{10} = 6a^5 \cdot 6a^5$. We factor such an expression by writing two factors that consist of the sum and the difference of the square roots of each of the squared terms. Thus,

$$25x^{12} - 36a^{10} = (5x^6 + 6a^5)(5x^6 - 6a^5).$$

Now do **C**.

Exercise 56   Factor completely $p^2 - 3p - 18$.

Solution   The expression is a trinomial with leading coefficient 1, so the factoring model is $x^2 + (a + b)x + ab = (x + a)(x + b)$. The first term is p^2, so

$$p^2 - 3p - 18 = (p + ?)(p + ?).$$

The last term is -18, so $a \cdot b = -18$. The coefficient of the middle term is -3, so $a + b = -3$. By a little trial and error, we determine that a and b are -6 and 3. Thus,

$$p^2 - 3p - 18 = (p - 6)(p + 3).$$

Now do **D**.

Exercise 64   Factor completely $2a^2 - 7ad + 6d^2$.

Solution   This expression illustrates the case of a general trinomial. The first term is $2a^2$, so

$$(2a^2 - 7ad + 6d^2) = (2a + ?)(a + ?).$$

The last term is $6d^2$ and the middle coefficient is negative, so the last terms in the factors are $-3d$ and $-2d$, or $-6d$ and $-d$. A little trial and error with various FOIL multiplications yields $(2a - 3d)(a - 2d)$ as the desired product.

Now do **E**.

E. Factor completely $2a^2 + 9ad - 5d^2$.

Exercise 80   Factor completely $(x - 2)^3 + 1$.

Solution   We use the factoring formula for the sum of two cubes with $a = x - 2$ and $b = 1$. Then

$$(x - 2)^3 + 1 = [(x - 2) + 1][(x - 2)^2 - (x - 2)(1) + 1^2]$$
$$= (x - 1)(x^2 - 4x + 4 - x + 2 + 1)$$
$$= (x - 1)(x^2 - 5x + 7).$$

Now do **F**.

F. Factor completely $(x + 2)^3 - 1$.

Exercise 94   Factor completely $x^3 - 3x^2 - x + 3$.

Solution   If we factor x^2 from the first two terms and -1 from the last two terms, we have $x^2(x - 3) - 1(x - 3)$. Then

$$x^2(x - 3) - 1(x - 3) = (x - 3)(x^2 - 1)$$
$$= (x - 3)(x + 1)(x - 1).$$

Now do **G**.

G. Factor completely $3x^3 - 12x^2 + 2x - 8$.

Exercise 98   Can $12s^2 - s + 1$ be factored with integer coefficients?

Solution   Matching $12s^2 - s + 1$ to the form $ax^2 + bx + c$ gives $a = 12$, $b = -1$, and $c = 1$. Then $b^2 - 4ac = (-1)^2 - 4(12)(1) = -47$. Since $b^2 - 4ac$ is not a perfect square, the expression cannot be factored (with integer coefficients).

Now do **H**.

H. Can $2x^2 - 3x - 3$ be factored with integer coefficients?

Exercise 102   Show that $\dfrac{(x + h)^3 - x^3}{h}$ simplifies to $3x^2 + 3xh + h^2$.

Solution $\dfrac{(x + h)^3 - x^3}{h} = \dfrac{(x^3 + 3x^2 h + 3xh^2 + h^3) - x^3}{h}$

$$= \dfrac{3x^2 h + 3xh^2 + h^3}{h} = \dfrac{h(3x^2 + 3xh + h^2)}{h}$$
$$= 3x^2 + 3xh + h^2$$

Now do **I**.

I. Simplify $\dfrac{(x + h)^4 - x^4}{h}$.

Exercise 108a   See text for question.

Solution   Acute angle A (opposite side a) and acute angle B (opposite side b) are complementary. Then since $A + B + \theta = 180°$, we know $\theta = 90°$. We now find the area of the figure in two ways. First, the figure is a square with side length $a + b$ so the area is $(a + b)^2$. Second, the figure is composed of a square with side length c and four right triangles each with area given by $\frac{1}{2}ab$. Thus,

$$(a + b)^2 = c^2 + 4[\tfrac{1}{2}ab]$$
$$a^2 + 2ab + b^2 = c^2 + 2ab$$
$$a^2 + b^2 = c^2.$$

1.4 Algebraic fractions

Objectives checklist

Can you:

a. Express a fraction in simplest form by using the fundamental principle?

b. Multiply two fractions and simplify the product?

c. Divide two fractions and simplify the quotient?

d. Add and subtract fractions and write the result in simplest form?

e. Change a complex fraction to a fraction in simplest form?

Key terms

Algebraic fractions   Complex fractions
Least common denominator (LCD)

Key rules and formulas

- *Equality principle:* $\dfrac{a}{b} = \dfrac{c}{d}$ if and only if $ad = bc$ $(b, d \neq 0)$

- *Fundamental principle:* $\dfrac{ak}{bk} = \dfrac{a}{b}$ $(b, k \neq 0)$

- *Simplification:* To express a fraction in simplest form:
    1. Factor completely the numerator and denominator of the fraction.
    2. Divide out nonzero factors that are common to the numerator and denominator according to the fundamental principle.

- *Multiplication:* The product of two or more fractions is the product of their numerators divided by the product of their denominators.

  In symbols the principle is $\dfrac{a}{b} \cdot \dfrac{c}{d} = \dfrac{ac}{bd}$ $(b, d \neq 0)$.

- *Division:* To divide two fractions, we invert the fraction by which we are dividing to find its reciprocal and then multiply.

  In symbols the principle is $\dfrac{a}{b} \div \dfrac{c}{d} = \dfrac{a}{b} \cdot \dfrac{d}{c}$ $\left(b, d, \dfrac{c}{d} \neq 0\right)$.

- *Addition (with common denominator):* The sum (or difference) of two or more fractions that have the same denominator is given by the sum (or difference) of the numerators divided by the common denominator.

  In symbols the principle is $\dfrac{a}{b} \pm \dfrac{c}{b} = \dfrac{a \pm c}{b}$ $(b \neq 0)$.

- *Addition (general procedure):* To add or subtract two or more fractions:
    1. Completely factor each denominator and find the LCD.
    2. For each fraction, obtain an equivalent fraction by applying the fundamental principle and multiplying the numerator and the denominator of the fraction by the factors of the LCD that are not contained in the denominator of that fraction.
    3. Add or subtract the numerators and divide this result by the common denominator.

Note: To add or subtract two fractions for which the LCD is simply the product of the denominators, apply the formula

$$\frac{a}{b} \pm \frac{c}{d} = \frac{ad \pm bc}{bd} \qquad (b, d \neq 0).$$

- *Least common denominator:* To find the LCD:

  1. Factor completely each denominator.
  2. The LCD is the product of all the different factors with each factor raised to the highest power to which it appears in any one of the factorizations.

- To simplify a complex fraction:

  1. Find the least common denominator of all the fractions that appear in the numerator and the denominator of the complex fraction.
  2. Multiply the numerator and the denominator of the complex fraction by this least common denominator and simplify your results.

Additional comments

- In addition $a + b = b + a$. However, in subtraction $a - b = -(b - a)$. Thus,

$$\frac{a + b}{b + a} = 1 \text{ while } \frac{a - b}{b - a} = -1.$$

Detailed solutions to selected exercises

Exercise 16 Express $\dfrac{z^3 - z^2 - 6z}{z^3 + z^2 - 12z}$ in simplest form.

Solution Proficiency in factoring and an understanding that only factors may be divided out are crucial here. Then

$$\frac{z^3 - z^2 - 6z}{z^3 + z^2 - 12z} = \frac{z(z^2 - z - 6)}{z(z^2 + z - 12)} = \frac{z(z - 3)(z + 2)}{z(z - 3)(z + 4)} = \frac{z + 2}{z + 4}.$$

Now do **A.**

Exercise 46 Express $\dfrac{2x^2 - 2ax + 2a^2}{(ax)^3} \div \dfrac{a^3 + x^3}{ax^3}$ in simplest form.

Solution If we factor each expression and write the division in terms of multiplication, we have

$$\frac{2(x^2 - ax + a^2)}{a^3 x^3} \cdot \frac{ax^3}{(a + x)(x^2 - ax + a^2)} = \frac{2}{a^2(a + x)}.$$

Now do **B.**

Exercise 68 Combine $\dfrac{16}{a^2 b} + \dfrac{1}{ab} - \dfrac{6}{ab^2}$ into a single fraction in simplest form.

Solution The least common denominator is $a^2 b^2$. Then

$$\frac{16}{a^2 b} + \frac{1}{ab} - \frac{6}{ab^2} = \frac{16(b)}{a^2 b(b)} + \frac{1(ab)}{ab(ab)} - \frac{6(a)}{ab^2(a)} = \frac{16b + ab - 6a}{a^2 b^2}.$$

Now do **C.**

1.4

A. Express $\dfrac{x^3 - x^2 - 12x}{x^3 - 8x^2 + 16x}$ in simplest form.

B. Express $\dfrac{x^2 + 3x + 2}{x} \div \dfrac{x^3 + 8}{x^2}$ in simplest form.

C. Combine $\dfrac{4}{xy} + \dfrac{2}{xy^2} - \dfrac{5}{yx^2}$ into a single fraction in simplest form.

D. Combine $\dfrac{3x}{x^2 - 9} - \dfrac{5 - x}{x + 3}$ into a single fraction in simplest form.

E. Simplify the expression $(x - 2)^{-1} - 3(x - 2)^{-3}$ into a single fraction in simplest form.

F. Change $\dfrac{x - \dfrac{1}{a}}{\dfrac{3}{a} + x}$ to a fraction in simplest form.

G. Change $\dfrac{\dfrac{x + h - 1}{x + h + 2} - \dfrac{x - 1}{x + 2}}{h}$ to a fraction in simplest form.

Exercise 80 Combine $\dfrac{2a}{a^2 - 1} - \dfrac{a + 1}{a - 1} + 7$ into a single fraction in simplest form.

Solution The least common denominator is $a^2 - 1$ or $(a + 1)(a - 1)$. We therefore rewrite the given expression as

$$\frac{2a}{a^2 - 1} - \frac{(a + 1)(a + 1)}{(a - 1)(a + 1)} + \frac{7(a^2 - 1)}{(a^2 - 1)} = \frac{2a - a^2 - 2a - 1 + 7a^2 - 7}{a^2 - 1}$$

$$= \frac{6a^2 - 8}{a^2 - 1}.$$

Now do **D.**

Exercise 94 Combine $(x - 4)^{-2} - 5(x - 4)^{-1}$ into a single fraction in simplest form.

Solution Rewrite the expression with positive exponents and combine as follows:

$$(x - 4)^{-2} - 5(x - 4)^{-1} = \frac{1}{(x - 4)^2} - \frac{5}{x - 4}$$

$$= \frac{1}{(x - 4)^2} - \frac{5(x - 4)}{(x - 4)(x - 4)}$$

$$= \frac{1 - 5x + 20}{(x - 4)^2}$$

$$= \frac{21 - 5x}{(x - 4)^2}.$$

Now do **E.**

Exercise 100 Change $\dfrac{x + (1/a)}{(2/a) - x}$ to a fraction in simplest form.

Solution The only denominator in the numerator and the denominator of the complex fraction is a, so

$$\frac{x + \dfrac{1}{a}}{\dfrac{2}{a} - x} = \frac{a\left(x + \dfrac{1}{a}\right)}{a\left(\dfrac{2}{a} - x\right)} = \frac{ax + 1}{2 - ax}.$$

Now do **F.**

Exercise 112 Change $\dfrac{\dfrac{x + h - 4}{x + h + 1} - \dfrac{x - 4}{x + 1}}{h}$ to a fraction in simplest form.

Solution If we multiply the numerator and the denominator of the complex fraction by $(x + h + 1)(x + 1)$, the original expression becomes

$$\frac{(x + h - 4)(x + 1) - (x + h + 1)(x - 4)}{h(x + h + 1)(x + 1)}$$

$$= \frac{x^2 + xh - 4x + x + h - 4 - x^2 - xh - x + 4x + 4h + 4}{h(x + h + 1)(x + 1)}$$

$$= \frac{5h}{h(x + h + 1)(x + 1)} = \frac{5}{(x + h + 1)(x + 1)}.$$

Now do **G.**

Exercise 116 Change $1 + (x + x^{-1})^{-1}$ to a fraction in simplest form.

H. Change $1 - (x - x^{-1})^{-1}$ to a fraction in simplest form.

Solution We rewrite the expression with positive exponents and proceed as follows:

$$
\begin{aligned}
1 + (x + x^{-1})^{-1} &= 1 + \frac{1}{x + (1/x)} \\
&= 1 + \left[\frac{1}{x + (1/x)} \right] \frac{x}{x} \\
&= 1 + \frac{x}{x^2 + 1} \\
&= \frac{x^2 + x + 1}{x^2 + 1}.
\end{aligned}
$$

Now do **H.**

1.5 Rational exponents

Objectives checklist

Can you:

a. Evaluate expressions containing rational number exponents?

b. Simplify expressions containing rational number exponents by using the laws of exponents?

c. Simplify expressions containing negative exponents by factoring out the smallest power of the variable from each term?

d. Convert an expression from rational exponent form to radical form, and vice versa?

Key terms

Radical
Radical sign
Radicand

Index of radical
Principal square root

Key rules and formulas

• For any positive integer n,

$$
a^{1/n} = \sqrt[n]{a} \begin{cases} \text{for } a \geq 0, \text{ if } n \text{ is even} \\ \text{for any real number } a, \text{ if } n \text{ is odd,} \end{cases}
$$

where

$$
\sqrt[n]{a} = b \text{ if and only if } b^n = a \begin{cases} \text{for } a \geq 0, b \geq 0, \text{ if } n \text{ is even} \\ \text{for any real number } a, \text{ if } n \text{ is odd.} \end{cases}
$$

• If m and n are integers with $n > 0$, and if m/n represents a reduced fraction such that $a^{1/n}$ represents a real number, then

$$
a^{m/n} = (a^{1/n})^m = (a^m)^{1/n}
$$

or equivalently

$$
a^{m/n} = (\sqrt[n]{a})^m = \sqrt[n]{a^m}.
$$

1.5

A. Simplify
$(1/3)^{-3} + 5(5)^0 - 25^{3/2}$.

B. Simplify $(3)^{6/5} \cdot (3)^{4/5}$.

C. Simplify $y/y^{1/3}$.

D. Simplify $\dfrac{4^0 x^{1/5} y^{-1/3}}{2^0 x^{-2} y^{-1/3}}$.

E. Simplify $\sqrt{x} \cdot \sqrt[n]{x}$.

Additional comments

- When n is even, $\sqrt[n]{}$ means the positive nth root.

- When n is even, the nth root of a negative number is not a real number.

- Unless the nth root of a number is a rational number, we leave results in radical form.

Detailed solutions to selected exercises

Exercise 38   Simplify $(\frac{1}{9})^{-2} + 2(3)^0 - 27^{2/3}$.

Solution   Term by term, we have $(\frac{1}{9})^{-2} = 9^2 = 81$, $2(3)^0 = 2 \cdot 1 = 2$, and $27^{2/3} = (\sqrt[3]{27})^2 = 3^2 = 9$. Thus, the given expression equals $81 + 2 - 9 = 74$.
Now do **A**.

Exercise 44   Simplify $(-4)^{7/5} \cdot (-4)^{8/5}$.

Solution   Our previous laws of exponents hold for exponents that are rational numbers. So, $(-4)^{7/5} \cdot (-4)^{8/5} = (-4)^{7/5 + 8/5} = (-4)^{15/5} = (-4)^3 = -64$.
Now do **B**.

Exercise 54   Simplify $\dfrac{x}{x^{1/2}}$.

Solution   To divide expressions with the same base, we subtract exponents as follows:

$$\frac{x}{x^{1/2}} = x^{1 - 1/2} = x^{1/2} \text{ or } \sqrt{x}.$$

Now do **C**.

Exercise 60   Simplify $\dfrac{2x^{4/5} y^{-2/3}}{6^0 x^{-1} y}$.

Solution   Note $6^0 = 1$ and subtract exponents as follows:

$$\frac{2x^{4/5} y^{-2/3}}{6^0 x^{-1} y} = 2x^{4/5 - (-1)} y^{-2/3 - 1} = 2x^{9/5} y^{-5/3} = \frac{2x^{9/5}}{y^{5/3}} \text{ or } \frac{2\sqrt[5]{x^9}}{\sqrt[3]{y^5}}.$$

Now do **D**.

Exercise 86   Simplify $\sqrt[m]{x} \cdot \sqrt[n]{x}$.

Solution   We convert from radical form to exponential form and use the laws of exponents to simplify the expression. So,

$$\sqrt[m]{x} \, \sqrt[n]{x} = x^{1/m} x^{1/n} = x^{1/m + 1/n} = x^{(m + n)/mn} \text{ or } \sqrt[mn]{x^{m + n}}.$$

Now do **E**.

Exercise 100   Simplify $\dfrac{1 + \frac{1}{2}(1 + x^2)^{-1/2}(2x)}{x + (1 + x^2)^{1/2}}$.

Solution   We multiply both the numerator and the denominator of the given fraction by $(1 + x^2)^{1/2}$ since $(1 + x^2)^{1/2} \cdot (1 + x^2)^{-1/2} = 1$.

$$\frac{1 + \frac{1}{2}(1 + x^2)^{-1/2}(2x)}{x + (1 + x^2)^{1/2}} \cdot \frac{(1 + x^2)^{1/2}}{(1 + x^2)^{1/2}} = \frac{(1 + x^2)^{1/2} + x \cdot 1}{[x + (1 + x^2)^{1/2}](1 + x^2)^{1/2}}$$

Now the expression $(1 + x^2)^{1/2} + x$ in the numerator is also a factor of the denominator so we simplify to $1/(1 + x^2)^{1/2}$ or $1/\sqrt{1 + x^2}$. Note in the simplification above that it is more useful to write the denominator in factored form than to multiply out the expression.

Now do **F**.

Exercise 104 Simplify $4x^{-1/3} + 12x^{-4/3}$ by factoring out the smallest power of the variable from each term.

Solution In this case $x^{-4/3}$ is the smallest power of x, so

$$4x^{-1/3} + 12x^{-4/3} = x^{-4/3}(4x + 12) = \frac{4x + 12}{x^{4/3}}.$$

Now do **G**.

F. Simplify $\dfrac{1 + (x - 1)^{-1}}{x(x - 1)^{1/2}}$.

G. Simplify $-2x^{-2/5} + 7x^{-3/5}$ by factoring out the smallest power of the variable from each term.

1.6 Operations with radicals

Objectives checklist

Can you:

a. Express a radical in simplest form?

b. Add and subtract radicals and simplify the result?

c. Multiply radicals and simplify the result?

d. Divide radicals and simplify the result?

e. Rationalize the numerator or the denominator of a fractional expression containing radicals?

f. Simplify expressions containing radicals by using combinations of the basic operations with radicals?

Key terms

Similar radicals

Conjugates

Rationalizing the denominator

Rationalizing the numerator

Key rules and formulas

- *Properties of radicals* (a, b, $\sqrt[n]{a}$, and $\sqrt[n]{b}$ denote real numbers)

  1. $(\sqrt[n]{a})^n = a$
  2. $\sqrt[n]{a} \cdot \sqrt[n]{b} = a^{1/n}b^{1/n} = (a \cdot b)^{1/n} = \sqrt[n]{a \cdot b}$
  3. $\dfrac{\sqrt[n]{a}}{\sqrt[n]{b}} = \dfrac{a^{1/n}}{b^{1/n}} = \left(\dfrac{a}{b}\right)^{1/n} = \sqrt[n]{\dfrac{a}{b}}$ $(b \neq 0)$

- To simplify a radical, do the following:

  1. Remove any factor of the radicand whose indicated root can be taken exactly.
  2. Eliminate any fractions in the radicand and all radicals from any denominators. (This procedure is called "rationalizing the denominator.")

- To add or subtract radicals, we first express each radical in simplest form and then we combine those that are similar by adding or subtracting their coefficients.

1.6

A. Simplify $\sqrt{18x^5y^{19}}$.

B. Simplify $\sqrt[3]{\dfrac{3x}{y^7}}$.

C. Simplify $4\sqrt{45} - 7\sqrt{27}$.

- To multiply expressions containing radicals, we use

$$\sqrt[n]{a} \cdot \sqrt[n]{b} = \sqrt[n]{ab} \quad \text{and} \quad (\sqrt[n]{a})^n = a$$

together with the normal rules of algebraic multiplication.

- To divide two expressions containing radicals, we use

$$\frac{\sqrt[n]{a}}{\sqrt[n]{b}} = \sqrt[n]{\frac{a}{b}} \qquad (b \neq 0)$$

together with the normal rules for algebraic division.

- If the denominator in a fraction contains square roots and is the *sum* of two terms, we rationalize the denominator by multiplying the numerator and the denominator of the fraction by its conjugate, the *difference* of the same two terms, and vice versa. To rationalize the numerator, follow the same procedure.

Additional comments

While $(\sqrt[n]{a})^n = a$ for all values of n, $\sqrt[n]{a^n} = |a|$ when n is even and $\sqrt[n]{a^n} = a$ when n is odd.

Detailed solutions to selected exercises

Exercise 8   Simplify $\sqrt{54x^{11}y^5}$ $(x, y > 0)$.

Solution   Rewrite $54x^{11}y^5$ as the product of a perfect square and another factor. Remember that any expression with an even exponent is a perfect square. Thus,

$$\sqrt{54x^{11}y^5} = \sqrt{9x^{10}y^4}\sqrt{6xy} = 3x^5y^2\sqrt{6xy}.$$

Now do **A**.

Exercise 18   Simplify $\sqrt[4]{2x/y^5}$ $(x, y > 0)$.

Solution   Rewrite $2x/y^5$ as an equivalent fraction whose denominator is a power that is a multiple of 4. So,

$$\sqrt[4]{\frac{2x}{y^5}} = \sqrt[4]{\frac{2x}{y^5} \cdot \frac{y^3}{y^3}} = \sqrt[4]{\frac{2xy^3}{y^8}} = \frac{\sqrt[4]{2xy^3}}{\sqrt[4]{y^8}} = \frac{\sqrt[4]{2xy^3}}{y^2}.$$

Now do **B**.

Exercise 26   Simplify $3\sqrt{28} - 10\sqrt{63}$.

Solution   First, express each radical in simplest form and then combine similar radicals.

$$3\sqrt{28} = 3\sqrt{4}\sqrt{7} = 3 \cdot 2\sqrt{7} = 6\sqrt{7}$$
$$10\sqrt{63} = 10\sqrt{9}\sqrt{7} = 10 \cdot 3\sqrt{7} = 30\sqrt{7}$$

Then $3\sqrt{28} - 10\sqrt{63} = 6\sqrt{7} - 30\sqrt{7} = -24\sqrt{7}$.
Now do **C**.

Exercise 48   Simplify $\sqrt[3]{54x^4y} - \sqrt[3]{xy^4/4}$.

Solution   First, simplify each radical.

$$\sqrt[3]{54x^4y} = \sqrt[3]{27x^3}\,\sqrt[3]{2xy} = 3x\sqrt[3]{2xy}$$

$$\sqrt[3]{\frac{xy^4}{4}} = \sqrt[3]{\frac{xy^4}{4} \cdot \frac{2}{2}} = \sqrt[3]{\frac{2xy^4}{8}} = \frac{\sqrt[3]{y^3}\sqrt[3]{2xy}}{\sqrt[3]{2^3}} = \frac{y\sqrt[3]{2xy}}{2}$$

Then

$$\sqrt[3]{54x^4y} - \sqrt[3]{xy^4/4} = 3x\sqrt[3]{2xy} - \frac{y}{2}\sqrt[3]{2xy} = \left(3x - \frac{y}{2}\right)\sqrt[3]{2xy}.$$

Now do **D**.

Exercise 50   Simplify $\sqrt{4 + 4x} + \sqrt{16 + 16x}$ $(x > 0)$.

Solution   Since $\sqrt{4 + 4x} = \sqrt{4}\sqrt{1 + x} = 2\sqrt{1 + x}$ and
$\sqrt{16 + 16x} = \sqrt{16}\sqrt{1 + x} = 4\sqrt{1 + x}$, we have
$\sqrt{4 + 4x} + \sqrt{16 + 16x} = 2\sqrt{1 + x} + 4\sqrt{1 + x} = 6\sqrt{1 + x}$.
Now do **E**.

Exercise 60   Multiply and simplify $(1 + \sqrt{2})^2$.

Solution   By definition, $(1 + \sqrt{2})^2 = (1 + \sqrt{2})(1 + \sqrt{2})$. We now multiply each term of the first factor by each term of the second factor and then simplify. Thus,

$$(1 + \sqrt{2})(1 + \sqrt{2}) = 1 + \sqrt{2} + \sqrt{2} + 2 = 3 + 2\sqrt{2}.$$

Note: $\sqrt{2} \cdot \sqrt{2} = 2$, and in general, $\sqrt{a} \cdot \sqrt{a} = a$ for all nonnegative values of a.
Now do **F**.

Exercise 64   Multiply and simplify $\sqrt{5x^3y} \cdot \sqrt{10y}$.

Solution $\sqrt{5x^3y} \cdot \sqrt{10y} = \sqrt{50x^3y^2}$, which simplifies to
$\sqrt{25x^2y^2} \cdot \sqrt{2x} = 5xy\sqrt{2x}$.
Now do **G**.

Exercise 68   Divide and simplify $-2\sqrt{54x^3y} \div 4\sqrt{3xy}$.

Solution   Writing the quotient as a fraction and using the properties of radicals, we have

$$\frac{-2\sqrt{54x^3y}}{4\sqrt{3xy}} = \frac{-2}{4}\sqrt{18x^2} = \frac{-2x}{4}\sqrt{18}, \text{ which simplifies to}$$

$$\frac{-\sqrt{18}x}{2} = \frac{-\sqrt{9}\sqrt{2}x}{2} = \frac{-3\sqrt{2}x}{2}.$$

Now do **H**.

Exercise 76   Rationalize the denominator of $\dfrac{\sqrt{3} + \sqrt{5}}{2\sqrt{3} - 7\sqrt{5}}$.

Solution   To rationalize $2\sqrt{3} - 7\sqrt{5}$, multiply it by its conjugate $2\sqrt{3} + 7\sqrt{5}$. To obtain an equivalent fraction, we must multiply by this number in both the numerator and the denominator. Thus,

$$\frac{\sqrt{3} + \sqrt{5}}{2\sqrt{3} - 7\sqrt{5}} = \frac{\sqrt{3} + \sqrt{5}}{2\sqrt{3} - 7\sqrt{5}} \cdot \frac{2\sqrt{3} + 7\sqrt{5}}{2\sqrt{3} + 7\sqrt{5}}$$

$$= \frac{6 + 9\sqrt{15} + 35}{12 - 245}$$

$$= \frac{41 + 9\sqrt{15}}{-233}$$

$$= \frac{-41 - 9\sqrt{15}}{233}.$$

Now do **I**.

D. Simplify $\sqrt[3]{27x^7} - x\sqrt[3]{x^4}$.

E. Simplify
$\sqrt[3]{8 - 8x} + \sqrt[3]{125 - 125x}$.

F. Multiply and simplify
$(2 - \sqrt{3})^2$.

G. Multiply and simplify
$\sqrt{3x^5y} \cdot \sqrt{12x^2y^3}$.

H. Divide and simplify
$6\sqrt{18xy^6} \div -2\sqrt{2xy^4}$.

I. Rationalize the denominator of
$\dfrac{3}{1 - \sqrt{5}}$.

J. Rationalize the numerator of $\dfrac{1 - \sqrt{x}}{1 + \sqrt{x}}$.

K. Simplify $\sqrt{\dfrac{y}{x}} \cdot \dfrac{x^2 \sqrt{\dfrac{x}{y}} - x}{y^2}$.

Exercise 84   Rationalize the numerator of $\dfrac{\sqrt{1 + h} - 1}{h}$.

Solution   To rationalize $\sqrt{1 + h} - 1$, multiply by its conjugate $\sqrt{1 + h} + 1$. So,

$$\frac{\sqrt{1 + h} - 1}{h} = \frac{\sqrt{1 + h} - 1}{h} \cdot \frac{\sqrt{1 + h} + 1}{\sqrt{1 + h} + 1}$$

$$= \frac{h}{h(\sqrt{1 + h} + 1)}$$

$$= \frac{1}{\sqrt{1 + h} + 1}.$$

Now do **J.**

Exercise 90   Simplify $-\dfrac{1}{2}\sqrt{\dfrac{x}{y}} \cdot \dfrac{x\sqrt{y/x} - y}{x^2}$.

Solution   Note $\sqrt{x/y} = \sqrt{xy}/y$ and $\sqrt{y/x} = \sqrt{xy}/x$. Then

$$-\frac{1}{2} \cdot \frac{\sqrt{xy}}{y} \cdot \frac{x(\sqrt{xy}/x) - y}{x^2} = \frac{xy - y\sqrt{xy}}{-2yx^2} = \frac{x - \sqrt{xy}}{-2x^2} = \frac{\sqrt{xy} - x}{2x^2}.$$

Now do **K.**

Sample test questions: Chapter 1

1. Evaluate.

   a. $\dfrac{8^{-1} - 5^0}{9^{-1/2} + 1^6}$ b. $\sqrt{\dfrac{1 + \frac{1}{8}}{5}}$ c. $\dfrac{1}{\sqrt{5}} + \dfrac{3\pi}{5}$

   d. $2 - [7 - (11 - 19)]$ e. $(4/3)^{-1}$ f. $(-64)^{2/3} + 27^{-1/3}$

   g. $\sqrt{(7\pi)^2 + (-\pi\sqrt{7})^2}$ h. $(\sqrt{18} - \sqrt{50})/2$

2. Evaluate each expression if $x = -2$, $y = 3$, and $z = -4$.

   a. $(-x)^{1/2} + (-z)^{-1/2} - y^0$ b. $3x^2 - x + 1$ c. $|4x| - |y|$

   d. $x - 2(y - 2z)$ e. $(x + y)/(y + z)$

3. If $x = -5$, find

   a. the absolute value of x b. the negative of x

   c. the reciprocal of x

4. Perform the indicated operations and simplify.

   a. $3 - 2[x - (2 - x)]$ b. $a^2b^3(2a - 3b + 4)$

   c. $\dfrac{(1/a) - 1}{(1/a) + 1}$ d. $a^{2x} \div a^x$

   e. $(xyz)^{-1}/(x^{-2}y)$ f. $(a - b)^{1/2} \div (a - b)^{1/4}$

   g. $\dfrac{3^{n+2}}{2^n} \div \dfrac{3^n}{(n + 1)2^{n+1}}$ h. $(a^{1/3} + a^{-1/3})^3$

   i. $\dfrac{(y^{-3/4}y^{3/8})^{-1}}{y^{1/4}}$ j. $x^2 - \dfrac{ax^2}{3x + a}$

k. $(2x + 3)(3 + x - x^2)$

l. $\dfrac{(x + h)^3 - x^3}{h}$

m. $\dfrac{(2 - a)x - (a - 2)y}{(a - 2)xy}$

n. $\dfrac{n^2 - n - 12}{n^2} \cdot \dfrac{n}{n - 4}$

o. $\dfrac{2x}{5} - \dfrac{7}{3x} + \dfrac{2}{x}$

p. $\dfrac{2 - t}{3 - 2t} - \dfrac{4}{2t - 3}$

q. $\left(\dfrac{y^2}{x^2} - 1\right) \div \left(\dfrac{y}{x} - 1\right)$

r. $(x + y)(x^2 - xy + y^2)$

5. Factor completely.

   a. $x^2 - 9x + 20$ b. $9a^2x^2 + 3ax$ c. $n^3 - n$

   d. $3k^2 - 6k - 24$ e. $8x^3 - y^3$ f. $x^3 - x^2 - 4x + 4$

6. True or False

   a. $\frac{22}{7}$ is an irrational number.

   b. The statement $a(b + c) = (b + c)a$ illustrates the distributive property.

   c. $-0.01 < -0.001$

   d. $|a| = a$ for all values of a.

7. Complete the statement.

   a. The least common denominator of $\dfrac{1}{x^2 + 5x + 6}$ and $\dfrac{5}{x^2 + 2x - 3}$ is _____ .

   b. The expanded form of $(\sqrt{x} + \sqrt{y})^2$ is _____ .

   c. The expression $(x - 2)/(2 - x)$ simplifies to _____ .

   d. As a single fraction, $a^{-1} + b^{-1}$ equals _____ .

   e. Rationalizing the denominator, we write $\dfrac{\sqrt{5} - \sqrt{2}}{\sqrt{5}}$ as _____ .

   f. Rationalizing the numerator, we write $\dfrac{\sqrt{x + 3}}{\sqrt{x^2 - 9}}$ as _____ .

8. Select the choice that completes the statement or answers the question.

   a. Which number is an irrational number?

      (a) $\sqrt{9}$ (b) $\sqrt{10}$ (c) $\sqrt{-1}$ (d) 0 (e) $3.1232323\ldots$

   b. Which statement illustrates the commutative property of addition?

      (a) $2(x + 5) = 2x + 2 \cdot 5$

      (b) $2(x + 5) = (x + 5) \cdot 2$

      (c) $(x + 5)/2 = x/2 + 5/2$

      (d) $2 + (x + 5) = 2 + (5 + x)$

      (e) $2 + (x + 5) = (2 + x) + 5$

c. If $a \cdot b = 1$, then a and b are called

   (a) negatives of each other   (b) opposites of each other

   (c) reciprocals of each other   (d) terms of each other

   (e) powers of each other

d. The side of a square is given by $3x^4$. The area of the square is

   (a) $9x^8$ (b) $3x^{16}$ (c) $3x^8$ (d) $6x^{16}$ (e) $9x^{16}$

e. Replace the question mark with the correct expression:
$$\frac{3 - a}{b + 1} = \frac{?}{1 + b}.$$

   (a) $3 + a$ (b) $a - 3$ (c) $3 - a$

   (d) $-a - 3$ (e) $-3 + a$

f. The expression $x^{-1/2}$ is equivalent to

   (a) $\sqrt{x}$ (b) $-\sqrt{x}$ (c) $-\sqrt{x}/2$ (d) $-1/\sqrt{x}$ (e) $1/\sqrt{x}$

g. If $x = 2 - \sqrt{7}$, then x^2 equals

   (a) -5 (b) -3 (c) -45 (d) -47 (e) $11 - 4\sqrt{7}$

h. Which statement is true for all values of x and y for which the statements are defined?

   (a) $\dfrac{2x + y}{x + y} = 2$ (b) $\dfrac{1}{x} + \dfrac{1}{y} = \dfrac{2}{xy}$

   (c) $2^x \cdot 2^y = 4^{xy}$ (d) $(x + 1)^2 = x^2 + 1$

   (e) $\dfrac{7y - xy}{y} = 7 - x$

i. Which statement is true for all values of x for which the statements are defined?

   (a) $\sqrt{x} + \sqrt{x} = \sqrt{2x}$ (b) $(\sqrt{x} + \sqrt{2})^2 = x + 2$

   (c) $\sqrt{x^2 + 1} = x + 1$ (d) $(\sqrt{x + 2})^2 = x + 2$

   (e) $\sqrt{x}\sqrt{x} = \sqrt{2x}$

Chapter 2
Equations and Inequalities

2.1 Equations and word problems

Objectives checklist

Can you:

a. Solve linear equations, or fractional equations that lead to linear equations, and check your answer?

b. Solve applied problems by setting up and solving an equation?

c. Solve a given formula for a specified variable?

Key terms

Equation   Root
Conditional equation   Ratio
Identity   Proportion
Equivalent equation   Empty set
First-degree (or linear) equation

Key rules and formulas

- *Equation-solving principles:* Two equations are equivalent when they have the same solution set. We change the original equation to an equivalent equation of the form $x =$ number by performing any combination of the following steps:

    1. Adding or subtracting the same expression to (from) both sides of the equation
    2. Multiplying or dividing both sides of the equation by the same (nonzero) number

- To solve equations containing fractions, do the following:

    1. Find the least common denominator or LCD.
    2. Multiply both sides of the equation by the LCD and obtain solutions by solving the resulting equation.
    3. Check each solution by substituting it in the original equation.

- *Solving a proportion:* Write the ratios as fractions (with the unknown in a numerator) and use the principles for solving equations.

- Formulas are solved for a given variable by using the equation-solving principles outlined above.

- *Analyzing word problems:* Consult the textbook for guidelines for setting up word problems.

Additional comments

- With equations that contain fractions it is important to check solutions in the original equation. Whenever we multiply both sides of an equation by a *variable,* we obtain an equation which, although true for the original solutions, might have additional solutions that must be rejected.

Chapter 2

2.1

A. Solve $5x - 5 = 2x + 6$.

Detailed solutions to selected exercises

Exercise 8 Solve $3z - 8 = 13z - 9$.

Solution
$$3z - 8 = 13z - 9$$
$$-10z - 8 = -9 \qquad \text{Subtract } 13z \text{ from both sides.}$$
$$-10z = -1 \qquad \text{Add 8 to both sides.}$$
$$z = \tfrac{1}{10} \qquad \text{Divide both sides by } -10.$$

The solution set is $\{\tfrac{1}{10}\}$.
Now do **A**.

B. Solve $3(x - 4) + 5 = 3x - 7$.

Exercise 16 Solve $2(x + 3) - 1 = 2x + 5$.

Solution
$$2(x + 3) - 1 = 2x + 5$$
$$2x + 6 - 1 = 2x + 5$$
$$2x + 5 = 2x + 5$$

Since $2x + 5 = 2x + 5$ is a true statement no matter which number we substitute for x, the original equation is true for all real numbers and is called an identity.
Now do **B**.

C. Solve $\dfrac{-2z - 16}{4} = 3$.

Exercise 26 Solve $\dfrac{2y - 7}{9} = 5$.

Solution $\dfrac{2y - 7}{9} = 5$
$$2y - 7 = 45 \qquad \text{Multiply both sides by 9.}$$
$$2y = 52 \qquad \text{Add 7 to both sides.}$$
$$y = 26 \qquad \text{Divide both sides by 2.}$$

The solution set is $\{26\}$.
Now do **C**.

D. Solve $\dfrac{x - 5}{x - 2} = \dfrac{x + 3}{x + 1}$.

Exercise 34 Solve $\dfrac{x + 4}{x + 1} = \dfrac{x + 2}{x + 3}$.

Solution The LCD is $(x + 1)(x + 3)$. Multiplying both sides of the equation by the LCD gives
$$(x + 3)(x + 4) = (x + 2)(x + 1)$$
$$x^2 + 7x + 12 = x^2 + 3x + 2$$
$$4x = -10$$
$$x = -\tfrac{10}{4} \text{ or } -\tfrac{5}{2}.$$

The solution, $-\tfrac{5}{2}$, checks in the original equation. The solution set is $\{-\tfrac{5}{2}\}$.
Now do **D**.

Exercise 46 The length of a rectangle is three times its width. If the perimeter is 160 in., determine the area of the rectangle.

Solution First, determine the length and the width. Let w represent the width of the rectangle. Then $3w$ represents the length of the rectangle. For a rectangle,
$$\text{perimeter} = 2(\text{length}) + 2(\text{width}).$$
$$\text{So, } 160 = 2(3w) + 2w$$
$$160 = 6w + 2w$$
$$160 = 8w$$
$$20 = w \quad \text{and so } 3w = 60.$$

The width is 20 in. and the length is 60 in., so the area is
$(20)(60) = 1,200$ sq in.
Now do **E**.

Exercise 50   One card sorter can process a deck of punched cards in 1 hour; another can sort the deck in 2 hours. How long would it take the two sorters together to process the cards?

Solution   Let x represent the number of hours required to process the cards with both sorters operating. (Note that x will be less than 1 whole hour.) In x hours the sorter that processes a deck in 1 hour will process x decks. The sorter that processes a deck in 2 hours processes $\frac{1}{2}$ a deck in 1 hour. In x hours it processes $\frac{1}{2}x$ of a deck. To complete the job, $x + \frac{1}{2}x$ must equal 1 deck. Solving this equation gives

$$
\begin{aligned}
x + \tfrac{1}{2}x &= 1 \qquad &\text{LCD: 2} \\
2x + 1x &= 2 \qquad &\text{Multiply both sides by the LCD.} \\
3x &= 2 \\
x &= \tfrac{2}{3}
\end{aligned}
$$

It will take $\frac{2}{3}$ of an hour, or 40 minutes, to process the cards.
Now do **F**.

Exercise 76   Solve $S = \dfrac{a}{1 - r}$ for r.

Solution $S = \dfrac{a}{1 - r}$

$$
\begin{aligned}
S(1 - r) &= a \qquad &\text{Multiply both sides by } (1 - r). \\
S - Sr &= a \qquad &\text{Use the distributive property.} \\
-Sr &= a - S \qquad &\text{Subtract } S \text{ from both sides.} \\
r &= \frac{a - S}{-S} \qquad &\text{Divide both sides by } -S. \\
\text{or } r &= \frac{S - a}{S} \qquad &\text{Multiply the numerator and the denominator by } -1.
\end{aligned}
$$

Now do **G**.

Exercise 86   Solve $\dfrac{1}{R} = \dfrac{1}{R_1} + \dfrac{1}{R_2}$ for R_2.

Solution   Multiplying both sides of the equation by $R \cdot R_1 \cdot R_2$ gives

$$
\begin{aligned}
R_1 R_2 &= R R_2 + R R_1 \\
R_1 R_2 - R R_2 &= R R_1 \qquad &\text{Subtract } R R_2 \text{ from both sides.} \\
R_2(R_1 - R) &= R R_1 \qquad &\text{Factor the common factor } R_2. \\
R_2 &= \frac{R R_1}{R_1 - R}. \qquad &\text{Divide both sides by } R_1 - R.
\end{aligned}
$$

Now do **H**.

2.2 Complex numbers

Objectives checklist

Can you:

a. Simplify an expression containing the square root of a negative number and write the result in the form $a + bi$?

E. The length of a rectangle is 6 in. more than its width. If the perimeter is 176 in., determine the area of the rectangle.

F. One painter can paint a room in 2 hours. Another can paint it in 3 hours. How long will it take the two painters working together to paint the room?

G. Solve $a = p(1 + rt)$ for r.

H. Solve $x = \dfrac{1}{y} + \dfrac{2}{z}$ for y.

b. Add, subtract, multiply, and divide complex numbers?

c. Determine if a given complex number is the solution of a specific equation?

d. Find the conjugate of a complex number?

e. Find the reciprocal of a complex number and write it in the form $a + bi$?

f. Verify the properties of conjugates for specific complex numbers?

g. Use the definition of conjugate to prove certain properties of conjugates?

Key terms

Complex number   Real part of a complex number
Imaginary number   Imaginary part of a complex number
Conjugate of a complex number

Key rules and formulas

- A number of the form $a + bi$, where a and b are real numbers and $i = \sqrt{-1}$, is called a complex number. The relationships among the various sets of numbers are shown below.

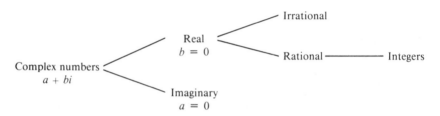

- Two complex numbers are equal if and only if their real parts are equal and their imaginary parts are equal.

$$a + bi = c + di \text{ if and only if } a = c \text{ and } b = d.$$

- The notation $\bar{z}$ denotes the conjugate of z. Thus, if $z = a + bi$, then $\bar{z} = a - bi$. The following are some of the basic properties of conjugates.

- *Properties of conjugates:* If z and w are complex numbers, then

  1. $z \cdot \bar{z}$ is a real number
  2. $\bar{z} = z$ if and only if z is a real number
  3. $\overline{z + w} = \bar{z} + \bar{w}$
  4. $\overline{z \cdot w} = \bar{z} \cdot \bar{w}$
  5. $\overline{z^n} = (\bar{z})^n$ for any positive integer n

- Informally, computations with complex numbers are similar to computations with real numbers that contain square roots. The important difference is that we must always express the imaginary part in terms of i, remembering that $i^2 = -1$. In formal terms the following rules are used to combine complex numbers.

  1. Two complex numbers are added by adding separately their real parts and their imaginary parts.

  $$(a + bi) + (c + di) = (a + c) + (b + d)i$$

  2. Two complex numbers are subtracted by subtracting separately their real parts and their imaginary parts.

  $$(a + bi) - (c + di) = (a - c) + (b - d)i$$

3. Two complex numbers are multiplied in the usual algebraic method with i^2 being replaced by -1.

$$(a + bi)(c + di) = ac + adi + bci + bdi^2$$
$$= (ac - bd) + (ad + bc)i$$

4. Two complex numbers are divided by multiplying the numerator and the denominator by the conjugate of the denominator.

$$\frac{a + bi}{c + di} = \frac{(a + bi)(c - di)}{(c + di)(c - di)} = \frac{(ac + bd) + (bc - ad)i}{c^2 + d^2}$$

Detailed solutions to selected exercises

Exercise 18 Simplify $\dfrac{-(-2) - \sqrt{(-2)^2 - 4(1)(5)}}{2(1)}$ in terms of i.

Solution First, $\sqrt{(-2)^2 - 4(1)(5)} = \sqrt{4 - 20} = \sqrt{-16} = 4i$. Then

$$\frac{-(-2) - \sqrt{(-2)^2 - 4(1)(5)}}{2(1)} = \frac{2 - 4i}{2}$$
$$= \frac{2}{2} - \frac{4i}{2} = 1 - 2i.$$

Now do **A**.

Exercise 34 Write the product $(2 - 5i)(3 - i)$ in the form $a + bi$.

Solution If we multiply each part of the first complex number by each part of the second complex number and replace i^2 by -1, we have

$$(2 - 5i)(3 - i) = 6 - 2i - 15i + 5i^2$$
$$= 6 - 2i - 15i + 5(-1) = 1 - 17i.$$

Now do **B**.

Exercise 50 Verify that $-4i$ is a solution of $x^4 + 13x^2 - 48 = 0$.

Solution We replace x by $-4i$ in the given equation and verify as follows:

$$(-4i)^4 + 13(-4i)^2 - 48 \overset{?}{=} 0$$
$$256 + 13(-16) - 48 \overset{?}{=} 0$$
$$256 - 208 - 48 \overset{?}{=} 0$$
$$0 \overset{\checkmark}{=} 0.$$

Note in the check above that $i^4 = i^2 \cdot i^2 = (-1)(-1) = 1$. Thus, $-4i$ is a solution of $x^4 + 13x^2 - 48 = 0$.
Now do **C**.

Exercise 52 Determine by checking if $2i$ is a solution of $x^2 - 4 = 0$.

Solution To check, we replace x by $2i$ in the given equation.

$$(2i)^2 - 4 \overset{?}{=} 0$$
$$4i^2 - 4 \overset{?}{=} 0$$
$$-4 - 4 \overset{?}{=} 0 \qquad \text{Note that } 4i^2 = -4 \text{ because } i^2 = -1.$$
$$-8 \neq 0$$

Therefore $2i$ is not a solution of the equation $x^2 - 4 = 0$.
Now do **D**.

2.2

A. Simplify
$$\frac{-4 - \sqrt{4^2 - 4(-2)(-3)}}{2(-2)}$$
in terms of i.

B. Write the product $(1 - 3i)(2 + i)$ in the form $a + bi$.

C. Verify that $3i$ is a solution of $x^4 + 2x^2 - 63 = 0$.

D. Determine by checking if $-3i$ is a solution of $x^2 + 9 = 0$.

E. Write the quotient $\dfrac{3i}{1-3i}$ in the form $a+bi$.

F. Find the reciprocal of $3+4i$ and write it in the form $a+bi$.

G. Verify that $\overline{z\cdot w}=\bar{z}\cdot\bar{w}$ for $z=2i$ and $w=1-3i$.

Exercise 66 Find the conjugate of the number 7.

Solution Since $7=7+0i$, its conjugate is $7-0i$, which is also equal to 7. Thus, the number 7 is its own conjugate. In general, a number is its own conjugate if and only if the number is a real number.

Exercise 72 Write the quotient $\dfrac{-2i}{6-2i}$ in the form $a+bi$.

Solution Two complex numbers are divided by multiplying the numerator and the denominator by the conjugate of the denominator. The conjugate of $6-2i$ is $6+2i$. Thus,

$$\frac{-2i}{6-2i}=\frac{-2i}{6-2i}\cdot\frac{6+2i}{6+2i}=\frac{-12i-4i^2}{36+12i-12i-4i^2}$$
$$=\frac{-12i-4(-1)}{36-4(-1)}=\frac{4-12i}{40}.$$

Then

$$\frac{4-12i}{40}=\frac{4}{40}-\frac{12i}{40}=\frac{1}{10}-\frac{3}{10}i.$$

Now do **E**.

Exercise 80 Find the reciprocal of $-2-2i$ and write the answer in the form $a+bi$.

Solution The reciprocal of $-2-2i$ is $\dfrac{1}{-2-2i}$. Since the conjugate of $-2-2i$ is $-2+2i$, we have

$$\frac{1}{-2-2i}=\frac{1}{-2-2i}\cdot\frac{-2+2i}{-2+2i}=\frac{-2+2i}{4-4i^2}$$
$$=\frac{-2+2i}{8}=\frac{-1}{4}+\frac{1}{4}i.$$

Thus, the reciprocal of $-2-2i$ is $-\frac{1}{4}+\frac{1}{4}i$ in standard form.
Now do **F**.

Exercise 82b Verify that $\overline{z\cdot w}=\bar{z}\cdot\bar{w}$ for $z=4-i$ and $w=5+2i$.

Solution We replace z by $4-i$ and w by $5+2i$ and verify
$$\overline{(4-i)(5+2i)}=\overline{(4-i)}\cdot\overline{(5+2i)}.$$
On the left side: $\overline{(4-i)(5+2i)}=\overline{20+8i-5i-2i^2}$
$$=\overline{22+3i}=22-3i.$$
On the right side: $\overline{(4-i)}\cdot\overline{(5+2i)}=(4+i)(5-2i)$
$$=20-8i+5i-2i^2=22-3i.$$
Thus, $\overline{z\cdot w}=\bar{z}\cdot\bar{w}$ for the given values of z and w.
Now do **G**.

Exercise 86 Use the definition of conjugate to prove that if $\bar{z}=z$, then z is a real number.

Solution Let $z=a+bi$, then $\bar{z}=a-bi$. If $z=\bar{z}$, then $a+bi=a-bi$, so by the definition of equality of complex numbers, $b=-b$, so $2b=0$, so $b=0$. Thus, $z=a+0i=a$ which shows that z is a real number.

2.3 Quadratic equations

Objectives checklist

Can you:

a. Solve quadratic equations by using the factoring method?

b. Solve quadratic equations by using the square root property?

c. Complete the square for an expression of the form $x^2 + bx$?

d. Solve a quadratic equation by completing the square?

e. Solve a quadratic equation by using the quadratic formula?

f. Use the discriminant to determine the nature of the solutions of a quadratic equation?

g. Solve applied problems involving quadratic equations?

Key terms

Quadratic (or second-degree) equation   Discriminant

Key rules and formulas

- *Zero product principle:* For any numbers a and b,

$$ab = 0 \text{ if and only if } a = 0 \text{ or } b = 0.$$

- We solve certain quadratic equations using factoring as follows:

  1. If necessary, change the form of the equation so that one side is 0.
  2. Factor the nonzero side of the equation.
  3. Set each factor equal to zero and obtain the solution(s) by solving the resulting equation.
  4. Check each solution by substituting it in the original equation.

- *Square root property:* If n is any real number, then

$$x^2 = n \text{ implies } x = \sqrt{n} \text{ or } x = -\sqrt{n}.$$

  This property may be used to solve quadratic equations in which there is no x term so the equation has the form $ax^2 + c = 0$.

- We complete the square for $x^2 + bx$ by adding $(b/2)^2$, which is the square of one half of the coefficient of x.

- *Quadratic formula:* If $ax^2 + bx + c = 0$ and $a \neq 0$, then

$$x = \frac{-b \pm \sqrt{b^2 - 4ac}}{2a}.$$

-

When a, b, c Are Rational and	The Solution(s) of $ax^2 + bx + c = 0$ Are
$b^2 - 4ac < 0$	conjugate complex numbers
$b^2 - 4ac = 0$	real, rational, equal numbers
$b^2 - 4ac > 0$ and a perfect square	real, rational, unequal numbers
$b^2 - 4ac > 0$ and not a perfect square	real, irrational, unequal numbers

2.3

A. Solve $3x^2 = -14x + 5$ by using the factoring method.

Detailed solutions to selected exercises

Exercise 14 Solve $2x^2 = 7x + 4$ by using the factoring method.

Solution First, rewrite the equation as $2x^2 - 7x - 4 = 0$. Factoring the left side of the equation gives $(2x + 1)(x - 4) = 0$. Now set each factor equal to zero. $2x + 1 = 0$ when $x = -\frac{1}{2}$, and $x - 4 = 0$ when $x = 4$. Thus, $-\frac{1}{2}$ and 4 are the solutions of the equation, and the solution set is $\{-\frac{1}{2}, 4\}$.
Now do **A.**

Exercise 24 Solve $(x + 3)^2 = 100$ by using the square root property.

Solution By the square root property, $(x + 3)^2 = 100$ implies $x + 3 = \pm 10$, so $x = -3 \pm 10$. Thus, the solution set is $\{-13, 7\}$.
Now do **B.**

B. Solve $(x - 2)^2 = 225$ by using the square root property.

Exercise 32 To complete the square for $x^2 + \frac{7}{3}x$, what number must be added?

Solution We complete the square by adding $(b/2)^2$. Since $b = \frac{7}{3}$, $b/2 = \frac{7}{6}$ and $(b/2)^2 = \frac{49}{36}$. Thus, to complete the square, we add $\frac{49}{36}$.
Now do **C.**

Exercise 38 Solve $5x^2 - 3x - 4 = 0$ by completing the square.

C. To complete the square for $x^2 - \frac{5}{2}x$, what number must be added?

Solution First, divide both sides of the equation by 5 (so the coefficient of x^2 is 1). Then write the x terms to the left of the equals and the constant to the right giving

$$x^2 - \tfrac{3}{5}x = \tfrac{4}{5}.$$

Now complete the square. Half of $-\frac{3}{5}$ is $-\frac{3}{10}$ and $(-\frac{3}{10})^2 = \frac{9}{100}$. Add $\frac{9}{100}$ to both sides of the equation and proceed as follows:

$$x^2 - \frac{3}{5}x + \frac{9}{100} = \frac{4}{5} + \frac{9}{100}$$

$$\left(x - \frac{3}{10}\right)^2 = \frac{89}{100}$$

$$x - \frac{3}{10} = \pm\sqrt{\frac{89}{100}}$$

$$x = \frac{3}{10} \pm \sqrt{\frac{89}{100}} = \frac{3 \pm \sqrt{89}}{10}.$$

D. Solve $3x^2 - 2x - 4 = 0$ by completing the square.

The solution set is $\left\{\dfrac{3 \pm \sqrt{89}}{10}\right\}$.

Now do **D.**

E. Solve $2x^2 = 5x + 5$ by using the quadratic formula.

Exercise 50 Solve $4x^2 = 12x - 9$ by using the quadratic formula.

Solution Expressing the equation in the form $ax^2 + bx + c = 0$, we write $4x^2 - 12x + 9 = 0$, so $a = 4$, $b = -12$, and $c = 9$. Then

$$x = \frac{-b \pm \sqrt{b^2 - 4ac}}{2a} = \frac{-(-12) \pm \sqrt{(-12)^2 - 4(4)(9)}}{2(4)}$$

$$= \frac{12 \pm \sqrt{144 - 144}}{8} = \frac{12 \pm \sqrt{0}}{8}.$$

Thus, $x_1 = (12 + 0)/8 = \frac{3}{2}$ and $x_2 = (12 - 0)/8 = \frac{3}{2}$. The roots x_1 and x_2 are equal and the solution set is $\{\frac{3}{2}\}$.
Now do **E.**

Exercise 62 Use the discriminant to determine the nature of the solutions of $-2x^2 + 4x + 9 = 0$.

Solution For the equation $-2x^2 + 4x + 9 = 0$, the discriminant $b^2 - 4ac = 4^2 - 4(-2)(9) = 88$. Since 88 is greater than 0 and not a perfect square, there are two different solutions that are irrational numbers.
Now do **F**.

F. Use the discriminant to determine the nature of the solutions of $3x^2 - 2x - 1 = 0$.

Exercise 64 The perimeter of a rectangle is 100 ft and the area is 400 ft². Find the dimensions of the rectangle.

Solution Since the perimeter is 100, we have

$$2\ell + 2w = 100 \quad \text{or} \quad \ell + w = 50.$$

Since the area is 400 we also know $\ell \cdot w = 400$. Substituting $50 - w$ for ℓ in the area formula gives

$$(50 - w)w = 400$$
$$0 = w^2 - 50w + 400.$$

Then by the quadratic formula,

$$w = \frac{-(-50) \pm \sqrt{(-50)^2 - 4(1)(400)}}{2(1)} = \frac{50 \pm \sqrt{2{,}500 - 1{,}600}}{2}$$
$$= \frac{50 \pm 30}{2} = 40 \text{ or } 10.$$

Thus, the dimensions are 40 ft by 10 ft.
Now do **G**.

G. The perimeter of a rectangle is 14 ft and the area is 12 ft². Find the dimensions of the rectangle.

2.4 Other types of equations

Objectives checklist

Can you:

a. Solve a radical equation and check your answer?

b. Solve an equation with quadratic form and check your answer?

c. Use factoring to solve certain higher degree polynomial equations?

Key terms

Radical equation   Equations with quadratic form
Extraneous solution

Key rules and formulas

- *Principle of powers:* If P and Q are algebraic expressions, then the solution set of the equation $P = Q$ is a subset of the solution set of $P^n = Q^n$, for any positive integer n.

- According to the principle of powers, every solution of $P = Q$ is a solution of $P^n = Q^n$; but solutions of $P^n = Q^n$ may or may not be solutions of $P = Q$. Solutions of $P^n = Q^n$ that do not satisfy the original equation are called extraneous solutions.

- An equation in which the unknown appears under a radical is called a radical equation. To solve such equations, do the following:

  1. Isolate a radical on one side of the equation.
  2. Raise both sides of the equation to the power that matches the index of the radical.

2.4

A. Solve
$$\sqrt[4]{x^2 + 4x + 4} = \sqrt[4]{x^2 + 6x - 8}.$$

B. Solve $\sqrt{x + 2} - 2 = x$.

3. Solve the resulting equation and then check all solutions in the *original* equation.

The check in Step 3 is essential to eliminate extraneous solutions.

- Equations that are not themselves quadratic but are equivalent to equations having the form $at^2 + bt + c = 0$ with $a \neq 0$ are called equations with quadratic form. To spot equations with quadratic form, look for the exponent in one term to be double the exponent in another term.

- To solve equations with quadratic form, we first let t equal a power of x that results in a quadratic equation. Then solve the quadratic equation for t, and finally solve the original equation for x.

- It is sometimes possible to factor higher degree polynomials into linear and/or quadratic factors with integer coefficients so that the equations can be solved by using the zero product principle.

Detailed solutions to selected exercises

Exercise 8   Solve $\sqrt[4]{x^2 + x + 6} = \sqrt[4]{x^2 + 3x + 2}$.

Solution   Raising both sides of the equation to the fourth power, we have

$$x^2 + x + 6 = x^2 + 3x + 2$$
$$x + 6 = 3x + 2$$
$$-2x = -4$$
$$x = 2.$$

Check: $\sqrt[4]{2^2 + 2 + 6} = \sqrt[4]{12}$, while $\sqrt[4]{2^2 + 3(2) + 2} = \sqrt[4]{12}$, so 2 is the solution, and the solution set is $\{2\}$.
Now do **A.**

Exercise 14   Solve $\sqrt{x + 1} + 1 = x$.

Solution   First, isolate the radical, then square both sides of the equation.

$$\sqrt{x + 1} = x - 1$$
$$x + 1 = x^2 - 2x + 1 \qquad \text{(squaring both sides)}$$
$$0 = x^2 - 3x$$
$$0 = x(x - 3)$$

Then $x(x - 3) = 0$, when $x = 0$ and when $x = 3$.
Check: $\sqrt{0 + 1} + 1 = 0$ gives $1 + 1 = 0$ so 0 is not a solution.
$\sqrt{3 + 1} + 1 = 0$ gives $2 + 1 = 3$ so 3 is a solution.
Therefore, the solution of the equation is 3, so the solution set is $\{3\}$.
Now do **B.**

Exercise 40   Solve $x^{1/3} - 1 = 2x^{-1/3}$.

Solution   First, eliminate negative exponents by multiplying both sides of the equation by $x^{1/3}$. Then

$$x^{2/3} - x^{1/3} = 2$$
$$(x^{1/3})^2 - x^{1/3} - 2 = 0.$$

Now if we let $t = x^{1/3}$, we have

$$t^2 - t - 2 = 0$$
$$(t - 2)(t + 1) = 0.$$

Thus, $t = 2, t = -1$. Now substitute $x^{1/3}$ for t and solve for x.

$$x^{1/3} = 2 \quad \text{so} \quad x = 2^3 = 8$$
$$x^{1/3} = -1 \quad \text{so} \quad x = (-1)^3 = -1$$

Both 8 and -1 check in the original equation and are solutions, so the solution set is $\{-1, 8\}$.
Now do **C**.

Exercise 48   Solve $x^6 = 10x^4 - 9x^2$.

Solution   Rewrite the equation so that one side is 0 and then factor completely.

$$x^6 - 10x^4 + 9x^2 = 0$$
$$x^2(x^4 - 10x^2 + 9) = 0$$
$$x^2(x^2 - 9)(x^2 - 1) = 0$$
$$x^2 = 0 \quad \text{or} \quad x^2 = 9 \quad \text{or} \quad x^2 = 1$$

So $x = 0$, $x = \pm 3$, or $x = \pm 1$. These numbers all check in the original equation and are therefore solutions. The solution set is $\{0, -1, 1, -3, 3\}$.
Now do **D**.

C. Solve $x^{1/2} - 1 = 2x^{-1/2}$.

D. Solve $x^5 = 20x^3 - 64x$.

2.5 Inequalities

Objectives checklist

Can you:

a. Solve inequalities and graph the solution?

b. Write number intervals using both inequalities and interval notation?

c. Solve applied problems involving inequalities?

Key terms

Solution of an inequality
Solution set
Transitive property
Equivalent inequalities
Set-builder notation
Interval
Interval notation

Endpoints of an interval
Open interval
Closed interval
Half-open interval
Infinity symbol
Union of two sets
Zero point

Key rules and formulas

- *Properties of inequalities:* The following properties may be used in solving inequalities. Let a, b, and c be real numbers.

	Comment
1. If $a < b$, then $a + c < b + c$.	The sense of the inequality is preserved when the same number is added to (or subtracted from) both sides of an inequality.
2. If $a < b$ and $c > 0$, then $ac < bc$.	The sense of the inequality is preserved when both sides of an inequality are multiplied (or divided) by the same positive number.

3. If $a < b$ and $c < 0$, then $ac > bc$. — The sense of the inequality is reversed when both sides of an inequality are multiplied (or divided) by the same negative number.

4. If $a < b$ and $b < c$, then $a < c$. — This property is called the transitive property.

Similar properties may be stated for the other inequality signs.

- Sets of real numbers may be written using interval notation or set notation and may be graphed as summarized in the accompanying chart.

Type of Interval	Interval Notation	Set Notation	Graph
Open interval	(a,b)	$\{x: a < x < b\}$	
Closed interval	$[a,b]$	$\{x: a \le x \le b\}$	
Half-open interval	$[a,b)$	$\{x: a \le x < b\}$	
	$(a,b]$	$\{x: a < x \le b\}$	
Infinite interval	$(-\infty,a)$	$\{x: x < a\}$	
	$(-\infty,a]$	$\{x: x \le a\}$	
	(a,∞)	$\{x: x > a\}$	
	$[a,\infty)$	$\{x: x \ge a\}$	
	$(-\infty,\infty)$	$\{x: x \text{ is a real number}\}$	

Note that ∞ and $-\infty$ are not real numbers but convenient symbols that help us designate intervals that are unbounded in the positive or negative direction.

- *Sign rule:* A product or quotient of nonzero factors is positive provided the number of negative factors is even. The product or quotient is negative if the number of negative factors is odd.

- We solve certain types of inequalities using factoring as follows:

  1. If necessary, change the form of the inequality so that one side is 0.
  2. Factor the nonzero side of the inequality.
  3. Make up a chart showing the signs of the factors with emphasis on the zero points.
  4. Apply the sign rule to determine the solution.

Additional comments

- When combining a pair of inequalities, it is important to distinguish between the mathematical meanings of "and" and "or." In an "and" statement both inequalities must be true simultaneously, whereas an "or" statement is true if at least one of the inequalities is true. Only an "and" statement can be written in compact form. To use interval notation for the set of numbers belonging either to set A or set B, we write $A \cup B$.

Detailed solutions to selected exercises

Exercise 6 Solve $16x - 7 \le 17x - 4$ and graph the solution.

Solution $16x - 7 \leq 17x - 4$

$-x - 7 \leq -4$ Subtract $17x$ from both sides.

$-x \leq 3$ Add 7 to both sides.

$x \geq -3$ Divide both sides by -3 and change the sense of the inequality.

Thus, the solution set is the interval $[-3,\infty)$. The graph of the solution set is shown in Figure 2.5:6.

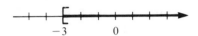

Figure 2.5:6

Now do **A**.

Exercise 20   Solve $7 - 4y \leq 4(7 - y)$ and graph the solution.

Solution $7 - 4y \leq 4(7 - y)$

$7 - 4y \leq 28 - 4y$ Use the distributive property.

$7 \leq 28$ Add $4y$ to both sides.

The resulting inequality is always true. Thus, the original inequality is true when y is replaced by any real number, and the graph is the real number line. Now do **B**.

Exercise 28   Write the set of numbers $\{x: -1 < x \leq 1\}$ in interval notation and draw its graph.

Solution   The interval excludes the left-hand endpoint -1 and includes the right-hand endpoint 1, so in interval notation we write $(-1,1]$. The graph is shown in Figure 2.5:28.

Figure 2.5:28

Now do **C**.

Exercise 38   Solve $-0.1 \leq 1 - x \leq 0.1$ and write the solution set in interval notation.

Solution   Using the properties of inequalities, we proceed as follows:

$-0.1 \leq 1 - x \leq 0.1$

$-1.1 \leq -x \leq -0.9$ Subtract 1 from each member.

$1.1 \geq x \geq 0.9$.   Divide each member by -1 and change the sense of the inequalities.

Thus, the solution set is the interval $[0.9,1.1]$.
Now do **D**.

Exercise 60   Solve $10 - c^2 > -3c$.

Solution   First, rewrite the inequality as $0 > c^2 - 3c - 10$. Factoring the right side of the inequality gives $0 > (c - 5)(c + 2)$. Now make up the chart below by placing zero points at 5 and -2 and applying the sign rule.

2.5

A. Solve $11x + 8 \leq -4x - 7$ and graph the solution.

B. Solve $16x - 8 \leq 8(2x + 3)$ and graph the solution.

C. Write the set of numbers $\{x: 1 \leq x < 4\}$ in interval notation and draw its graph.

D. Solve $-5 \leq -2x + 1 \leq 5$ and write the solution set in interval notation.

E. Solve $x^2 > 12 - 4x$.

F. Solve $\dfrac{1 + 2x}{3x + 2} > 1$.

G. To the nearest inch, the height of a man is 73 in. Use interval notation to write an interval that contains the man's exact height.

Sign of $c - 5$	$-$		$-$	0	$+$
				5	

Sign of $c + 2$	$-$	0	$+$		$+$
		-2			

Sign of $(c - 5)(c + 2)$	$+$	0	$-$	0	$+$
		-2		5	

As shown in the chart, $(c - 5)(c + 2)$ is negative if $-2 < c < 5$, so the solution set is $(-2, 5)$.
Now do **E**.

Exercise 80 Solve $\dfrac{2}{x - 1} < \dfrac{1}{x + 1}$ and write the solution set in interval notation.

Solution First, rewrite the inequality so the right side is zero. Then simplify the resulting expression on the left side into a single fraction.

$$\frac{2}{x - 1} < \frac{1}{x + 1} \quad \text{so} \quad \frac{2}{x - 1} - \frac{1}{x + 1} < 0 \quad \text{so} \quad \frac{x + 3}{(x - 1)(x + 1)} < 0$$

Now use the sign rule to make up the chart below.

Sign of $x + 3$	$-$	0	$+$	$+$		$+$
		-3				

Sign of $x - 1$	$-$		$-$	$-$	0	$+$
					1	

Sign of $x + 1$	$-$		$-$	0	$+$	$+$
			-1			

Sign of $\dfrac{x + 3}{(x - 1)(x + 1)}$	$-$	0	$+$	Undefined	$-$	Undefined	$+$
		-3		-1		1	

From the chart we read that the expression is less than zero if $x < -3$ or $-1 < x < 1$, so the solution set is $(-\infty, -3) \cup (-1, 1)$.
Now do **F**.

Exercise 82 See text for question.

Solution If the height of the woman is 69 in., to the nearest inch, we can guarantee that her exact height is somewhere in the interval [68.5 in., 69.5 in.). If her height is measured to be 69.3 in., then we can guarantee that her exact height is somewhere in the interval [69.25 in., 69.35 in.).
Now do **G**.

Exercise 88 The length of a rectangle is 2 ft more than the width, and the area is less than 63 ft². What are the possibilities for the length in such a case?

Solution Let ℓ represent the length of the rectangle. Then $\ell - 2$ represents the width of the rectangle. For a rectangle, area = length · width, so

$$\ell(\ell - 2) < 63.$$

We solve this inequality by first making the right side equal zero and then factoring the left side.

$$\ell(\ell - 2) - 63 < 0$$
$$\ell^2 - 2\ell - 63 < 0$$
$$(\ell - 9)(\ell + 7) < 0$$

Now make up a chart showing the zero points and apply the sign rule.

Sign of $\ell - 9$	$-$	$-$	$-$	0	$+$
				9	

Sign of $\ell + 7$	$-$	0	$+$	$+$	$+$
		-7			

Sign of $(\ell - 9)(\ell + 7)$	$+$	0	$-$	0	$+$
		-7		9	

We read from the chart that the expression will be negative for $-7 < \ell < 9$. However, the length is 2 ft more than the width and the width may not be negative or zero. Thus, ℓ may be any number in the interval (2 ft, 9 ft). Now do **H.**

2.6 Absolute value equations and inequalities

Objectives checklist

Can you:

a. Use the proper symbol ($<$, $>$, $=$) between expressions involving absolute value to indicate their correct order?

b. Solve equations containing absolute value?

c. Solve inequalities containing absolute value?

d. Write a number interval using inequalities and absolute value?

Key terms

Absolute value

Key rules and formulas

* $|a| = \begin{cases} a \text{ if } a \geq 0 \\ -a \text{ if } a < 0 \end{cases}$ (by definition)

*   Given any two points a and b, the distance between them on the number line is $|a - b|$. Note that $|a - b| = |b - a|$.

*   If $b > 0$, then $|a| < b$ is equivalent to $-b < a < b$.

*   Geometrically, $|a| < b$ says a is contained in an interval that begins and ends b units from 0.

*   If $b > 0$, then $|a| > b$ is equivalent to $a > b$ or $a < -b$.
    Geometrically, $|a| > b$ says a is located more than b units from zero.

* $|a + b| \leq |a| + |b|$
    $|a - b| \geq |a| - |b|$
    $|ab| = |a| \cdot |b|$
    $\left| \dfrac{a}{b} \right| = \dfrac{|a|}{|b|}$ $(b \neq 0)$

H. The width of a rectangle is 3 ft less than the length, and the area is less than 40 ft². What are the possibilities for the width in such a case?

2.6

A. Insert the proper symbol
$(>, <, =)$:

$|3(1 - x)|$ _____ $|3| \, |1 - x|$.

B. Solve $|-2x + 1| = x + 1$.

C. Solve $|x + b| < c$, $(c > 0)$.

D. Solve $|3 - x| + 5 \geq 13$.

E. Solve $|2x - 3| > x$.

Detailed solutions to selected exercises

Exercise 16   In the expression $|4(x - 1)|$ _____ $|4| \, |x - 1|$, insert the proper symbol $(<, >, =)$.

Solution   Since the absolute value of a product is the product of the absolute values (that is, $|ab| = |a| \cdot |b|$), we know $|4(x - 1)| = |4| \, |x - 1|$.
Now do **A.**

Exercise 36   Solve $|-x + 2| = x + 1$.

Solution   By definition, $|a|$ equals either a or $-a$. Thus, $|-x + 2|$ equals either $-x + 2$ or $-(-x + 2)$. Setting these expressions equal to $x + 1$, we have

$$
\begin{array}{ll}
-x + 2 = x + 1 \quad \text{or} & -(-x + 2) = x + 1 \\
{-2x} = -1 & x - 2 = x + 1 \\
x = \tfrac{1}{2} & \text{no solution.}
\end{array}
$$

Thus, $x = \tfrac{1}{2}$, and the solution set is $\{\tfrac{1}{2}\}$.
Now do **B.**

Exercise 46   Solve $|x - a| < d \quad (d > 0)$.

Solution   Using the rule that $|a| < b$ is equivalent to $-b < a < b$ in the stated problem gives

$$
\begin{aligned}
-d &< x - a < d \\
a - d &< x - a + a < a + d. \quad \text{Add } a \text{ to each component.}
\end{aligned}
$$

So $|x - a| < d$ is equivalent to $a - d < x < a + d$, so the solution set is $(a - d, a + d)$. Geometrically, $|x - a| < d$ says x is contained in an interval that begins and ends d units from a.
Now do **C.**

Exercise 56   Solve $|1 - x| + 2 \geq 4$.

Solution
$$
\begin{aligned}
|1 - x| + 2 &\geq 4 \\
|1 - x| &\geq 2
\end{aligned}
$$
We now use the rule that $|a| > b$ is equivalent to $a > b$ or $a < -b$ and proceed as follows:

$$
\begin{array}{lll}
1 - x \geq 2 & \text{or} & 1 - x \leq -2 \\
-x \geq 1 & & -x \leq -3 \\
x \leq -1 & & x \geq 3.
\end{array}
$$

The solution set is $(-\infty, -1] \cup [3, \infty)$.
Now do **D.**

Exercise 60   Solve $|3x - 7| > x$.

Solution   First, from the definition of absolute value it is obvious that $|3x - 7| > x$ is true when x is negative and when $x = 0$. To determine positive solutions, we use the rule that if $b > 0$, then $|a| > b$ is equivalent to $a > b$ or $a < -b$. Thus, if $x > 0$, then

$$3x - 7 > x \quad \text{or} \quad 3x - 7 < -x$$

which simplifies to

$$x > \tfrac{7}{2} \quad \text{or} \quad x < \tfrac{7}{4}.$$

Combining our results, the solution set is $(-\infty, \tfrac{7}{4}) \cup (\tfrac{7}{2}, \infty)$.
Now do **E.**

Exercise 68 Write the interval $(-\infty, -0.1) \cup (0.1, \infty)$, using inequalities and absolute value.

Solution The given interval consists of all real numbers that belong to either $(-\infty, -0.1)$ or $(0.1, \infty)$. This set may also be written as $\{x: x < -0.1$ or $x > 0.1\}$. By applying the rule that $|a| > b$ is equivalent to $a > b$ or $a < -b$, we have that $x > 0.1$ or $x < -0.1$ is equivalent to $|x| > 0.1$. So the given interval is equal to $\{x: |x| > 0.1\}$.
Now do **F**.

F. Write the interval given by $(-\infty, -5) \cup (5, \infty)$ using inequalities and absolute value.

Exercise 82 Use $|x| = \sqrt{x^2}$ to show that $|a + b| \le |a| + |b|$. (*Hint:* Note that $a^2 + 2ab + b^2 \le |a|^2 + 2|a| \cdot |b| + |b|^2$.)

Solution $|a + b|^2 = (a + b)^2 = a^2 + 2ab + b^2 \le |a|^2 + 2|a| \cdot |b| + |b|^2$
Since $|a|^2 + 2|a| \cdot |b| + |b|^2 = (|a| + |b|)^2$, we have

$$|a + b|^2 \le (|a| + |b|)^2$$
$$\text{so } \sqrt{|a + b|^2} \le \sqrt{(|a| + |b|)^2}$$
$$\text{so } |a + b| \le |a| + |b|.$$

Sample test questions: Chapter 2

1. Solve each equation.

   a. $\dfrac{|x|}{3} = |x| - 5$

   b. $\dfrac{1}{c^2 - c} = \dfrac{2}{c - 1}$

   c. $ax + bx = c$ (for x)

   d. $x^2 + 1 = 0$

   e. $\sqrt{2x + 1} - 7 = 0$

   f. $2x^2 = 2 - 5x$

   g. $x - \sqrt{x + 1} = 1$

   h. $(x + 1)^{1/3} = -2$

   i. $x^{-2} = 4x^{-1} + 3$

2. Solve each inequality. Write the solution set in interval notation.

   a. $|3x - 4| < 2$

   b. $x^2 - x < 12$

   c. $\dfrac{1}{x + 3} \ge \dfrac{2}{x - 1}$

   d. $3 - 4x > 0$

3. Perform the indicated operations and express the result in the form $a + bi$.

   a. $(6 + 2i) - (2 - 4i)$

   b. $(2 - 5i)(-2 + 3i)$

   c. $-4i \div (2 + i)$

4. If $z_1 = 1 - i$ and $z_2 = -\sqrt{2} - \sqrt{2}\, i$, then

   a. Find $z_1 \cdot z_2$ and express the result in the form $a + bi$.

   b. Find z_2^4 and express the result in the form $a + bi$.

5. Solve each applied problem.

   a. A person drives from her home to work at an average speed of 30 mph. What speed must she average along the same route on the way home for her average speed for the round trip to be 40 mph?

   b. The height (y) of a projectile that is shot directly up from the ground with an initial velocity of 80 ft/second is given by the formula $y = 80t - 16t^2$. When will the projectile strike the ground?

c. An oil storage tank is being filled through two pipes. One pipe working alone can fill the tank in 12 hours; the other pipe can fill the tank in 8 hours. How long will it take to fill the tank when both pipes are used?

d. The height (y) of a projectile that is shot directly up from the ground with an initial velocity of 100 ft/second is given by the formula $y = 100t - 16t^2$. To the nearest hundredth of a second, when will the projectile initially attain a height of 120 ft?

6. Complete the statement.

a. The conjugate of $-2 + 3i$ is _____ .

b. In terms of i, the number $\sqrt{-25}$ is written as _____ .

c. If $x = 1 + i$, then $x^2 - 2x + 3 =$ _____ .

d. The value of the discriminant in $3x^2 - 2x - 1 = 0$ is _____ .

e. To complete the square for $x^2 - 10x$, we add _____ .

7. Select the choice that completes the statement or answers the question.

a. If $\dfrac{a}{b} + \dfrac{b}{x} = b$, then x equals

    (a) $\dfrac{1 - a^2}{b}$ (b) $b - \dfrac{a}{b^2}$ (c) $\dfrac{b^2}{b^2 - a}$ (d) $1 - \dfrac{a}{b^2}$

b. If the width of a rectangle is one-third the length and the perimeter is 32, what is the area?

    (a) 4   (b) 48   (c) 24   (d) 12

c. Pick the statement that corresponds to the given figure.

    (a) $|x| < 2$ (b) $|x| \le 2$ (c) $|x| \ge 2$ (d) $|x| > 2$

d. If $a = b\sqrt{c/d}$, then c equals

    (a) $\dfrac{a^2d}{b^2}$ (b) $\dfrac{b}{ad^2}$ (c) $\left(\dfrac{ad}{b}\right)^2$ (d) $\dfrac{ab^2}{d}$

Chapter 3
Functions and Graphs

3.1 Functions

Objectives checklist

Can you:

a. Determine whether a given correspondence or set of ordered pairs is a function?

b. Find the domain and range of a function?

c. Determine ordered pairs that are solutions of an equation?

d. Find $f(x)$, given a rule f and an x value?

e. Find x, given a rule f and a y value (or $f(x)$)?

f. Write the difference quotient for a function in simplest form?

g. Find a formula that defines the functional relationship between two variables?

Key terms

Function   Independent variable
Domain   Dependent variable
Range   Functional notation
Ordered pairs   Constant function

Function: A function is a rule that assigns to each element x in a set X exactly one element y in a set Y. In this definition set X is called the domain of the function, and the set of all elements of Y that correspond to elements in the domain is called the range of the function.

Ordered pairs: In mathematical notation we represent the correspondence in a function by using ordered pairs. In such pairs the value for the independent variable is listed first and the value for the dependent variable comes second.

Alternate definition of a function: A function is a set of ordered pairs in which no two different ordered pairs have the same first component. The set of all first components (of the ordered pairs) is called the domain of the function. The set of all second components is called the range of the function.

Key rules and formulas

- The term $f(x)$ is read "f of x" or "f at x" and means the value of the function (the y value) corresponding to the value of x.

- If $y = f(x)$, and a is in the domain of f, then $f(a)$ means the value of y when $x = a$. Thus, evaluating $f(a)$ often requires nothing more than a substitution.

- $f(a)$ is a y value; a is an x value. Hence, ordered pairs for the function defined by $y = f(x)$ all have the form $(a, f(a))$.

- In functional notation we use the symbols f and x more out of custom than necessity and other symbols work just as well. The notations $f(x) = 2x$, $f(t) = 2t$, $g(y) = 2y$ and $h(z) = 2z$ all define exactly the same function if x, t, y, and z may be replaced by the same numbers.

Chapter 3

3.1

A. Does the equation $y^2 = x + 1$ determine y as a function of x?

B. Find the domain and range of the function $y = \dfrac{1}{x - 4}$.

C. Determine solutions of $y = -|x|$ from the ordered pairs $(-1,1)$, $(1,1)$, $(-1,-1)$, and $(1,-1)$.

D. If $y = -2x + 3$, determine the domain element that corresponds to a range element of 9.

E. Is $\{(2,1), (3,4), (4,1)\}$ a function?

Detailed solutions to selected exercises

Exercise 2 See text for question.

Solution Since each member of set X is assigned exactly one member of set Y, the correspondence is a function. It doesn't matter that two different values of x, \$5 and \$6, both have the same y value, \$170,000, assigned to them.

Exercise 6 Does the equation $x = y^4$ determine y as a function of x?

Solution Since $x = y^4$ implies $y = \pm \sqrt[4]{x}$, it is possible for an x value to correspond to two different y values. For example, if $x = 1$, then $y = 1$ or $y = -1$. Thus, $x = y^4$ does not determine y as a function of x.
Now do **A**.

Exercise 26 Find the domain and range of the function $y = 1/\sqrt{x + 5}$.

Solution We exclude from the domain values for the independent variable (x) that result in the square root of a negative number or in division by zero. Thus, for $y = 1/\sqrt{x + 5}$, we require $x + 5 > 0$, so $x > -5$, and the domain is the interval $(-5,\infty)$. If $x > -5$, then $1/\sqrt{x + 5}$ is greater than 0, so the range is the interval $(0,\infty)$.
Now do **B**.

Exercise 32 Determine solutions to $y = |x|$ from the ordered pairs $(-1,-1)$, $(1,1)$, $(-1,1)$, and $(1,-1)$.

Solution In the ordered pairs the x value is listed first and the y value comes second. Thus,

$-1 \overset{?}{=} |-1|$	$1 \overset{?}{=} |1|$	$1 \overset{?}{=} |-1|$	$-1 \overset{?}{=} |1|$
$-1 \neq 1$	$1 = 1$	$1 = 1$	$-1 \neq 1$
For $(-1,-1)$, No	For $(1,1)$, Yes	For $(-1,1)$, Yes	For $(1,-1)$, No

Now do **C**.

Exercise 40 If $y = -x - 1$, then determine the domain element that corresponds to a range element of 6.

Solution Setting $y = 6$ and solving for x gives $x = -7$. Thus, the domain element is -7.
Now do **D**.

Exercise 46 Is $\{(1,2), (2,2), (3,2)\}$ a function?

Solution Yes. The first component in the ordered pairs is always different.
Now do **E**.

Exercise 48 State the domain and range of $\{(0,5), (1,6), (2,7), (3,8)\}$.

Solution The collection of first components is the domain of the function, while the collection of second components is the range. Thus, the domain is $\{0,1,2,3\}$ and the range is $\{5,6,7,8\}$.
Now do **F.**

Exercise 50 If $f(t) = t^2 + 1$, find $f(-1)$, $f(0)$, and $f(1)$.

Solution Replace t by the given number and simplify.

$$f(-1) = (-1)^2 + 1 = 2$$
$$f(0) = 0^2 + 1 = 1$$
$$f(1) = 1^2 + 1 = 2$$

Now do **G.**

Exercise 56c If $g(x) = 2x + 7$, for what value of x will $g(x) = -3$?

Solution Replace $g(x)$ by -3 and solve the resulting equation.

$$-3 = 2x + 7$$
$$-5 = x$$

Now do **H.**

Exercise 60 If $g(x) = x^2$, find (a) $g(2)$, (b) $g(-2)$, (c) Does $g(x) = g(-x)$ for all values of x?

Solution (a) $g(2) = 2^2 = 4$, (b) $g(-2) = (-2)^2 = 4$, (c) Yes, since $g(-x) = (-x)^2 = x^2 = g(x)$.
Now do **I.**

Exercise 64 See text for question.

Solution
a. Since 3 is greater than 1, use $h(x) = -1$, so $h(3) = -1$.

b. Since 1 is in the interval $0 \le x \le 1$, use $h(x) = x$, so $h(1) = 1$.

c. Since $\frac{1}{2}$ is in the interval $0 \le x \le 1$, use $h(x) = x$, so $h(\frac{1}{2}) = \frac{1}{2}$.

d. Since 0 is in the interval $0 \le x \le 1$, use $h(x) = x$, so $h(0) = 0$.

e. Since -3 is not in the domain of the function, $h(-3)$ is undefined.

Now do **J.**

Exercise 68 Write the difference quotient, in simplest form, for the function $f(x) = 7$.

Solution Since $f(x) = 7$ assigns the same value, 7, to each value of x, we have $f(x + h) = 7$. Then if $h \ne 0$,

$$\frac{f(x + h) - f(x)}{h} = \frac{7 - 7}{h} = \frac{0}{h} = 0.$$

F. State the domain and range of the function $\{(5,3), (4,2), (3,1), (2,0)\}$.

G. If $f(t) = 2t^3 - 1$, find $f(-1)$, $f(0)$, and $f(1)$.

H. If $g(x) = 3x + 7$, for what value of x will $g(x) = -2$?

I. If $g(x) = 2x^2$, find (a) $g(2)$, (b) $g(-2)$, (c) Does $g(x) = g(-x)$ for all values of x?

J. If $f(x) = \begin{cases} 2x \text{ if } x > 3 \\ 5 \text{ if } x \le 3, \end{cases}$ find (a) $f(2)$, (b) $f(3)$, (c) $f(5)$.

K. Write the difference quotient, in simplest form, for the function $f(x) = x$.

Now do **K.**

Exercise 76   Express the area (A) of a square as a function of the perimeter (P) of the square.

Solution   The common formula for the area of a square is $A = s^2$. To answer the question, we need to express s in terms of P. The perimeter formula $P = 4s$ tells us $s = P/4$, so $A = s^2 = (P/4)^2$. Since the perimeter of a square must be positive, the domain is the interval $(0,\infty)$.
Now do **L.**

Exercise 90   See text for question.

L. Express the perimeter (P) of a square as a function of the area (A) of the square.

Solution   If the number (n) of units is 2 or less, the cost is $2.06 so

$$c = 2.06 \quad \text{if} \quad 0 \le n \le 2.$$

For the next 6 units, meaning $2 < n \le 8$, the cost is $2.06 plus 40 cents for each unit above 2. Thus,

$$c = 2.06 + 0.40(n - 2) \quad \text{if} \quad 2 < n \le 8.$$

Finally, if $8 < n \le 14$, the cost is $2.06 for the first 2 units, plus $2.40 for the next 6 units, plus $0.265(n - 8)$ for the units above 8. Thus, the cost function is

M. Express the cost (c) of having carpeting installed in terms of the number (n) of square yards being installed if the carpeting consists of no more than 100 square yards and the carpet installers have the following fee schedule:

$$c = \begin{cases} 2.06 & \text{if} \quad 0 \le n \le 2 \\ 2.06 + 0.40(n - 2) & \text{if} \quad 2 < n \le 8; \\ 4.46 + 0.265(n - 8) & \text{if} \quad 8 < n \le 14 \end{cases} \quad \text{Domain: } [0,14].$$

Now do **M.**

	Amount	Charge
First	20 sq yd or less	$150
Next	30 sq yd	$5/sq yd
Next	50 sq yd	$4/sq yd

3.2 Cartesian coordinates and graphs

Objectives checklist

Can you:

a. Graph an ordered pair?

b. Determine the coordinates of a given point?

c. Find the distance between two points?

d. Graph a function by finding and plotting ordered pairs in the function and connecting these points with the appropriate curve (which can be forecast in many cases from the form of the equation)?

e. Read from a given graph the domain, range, function values, and values of x for which $f(x) = 0$, $f(x) < 0$, and $f(x) > 0$?

f. Determine if a given graph is the graph of a function?

Key terms

Cartesian coordinate system
Axes
Origin

Quadrant
Coordinates
Graph

Key rules and formulas

*   *Distance formulas:*
    $d = |x_2 - x_1|$ (for two points with the same y-coordinate)
    $d = |y_2 - y_1|$ (for two points with the same x-coordinate)
    $d = \sqrt{(x_2 - x_1)^2 + (y_2 - y_1)^2}$ (for any two points)

- The fundamental principle in analytic geometry is that every ordered pair that satisfies an equation corresponds to a point in its graph and every point in the graph corresponds to an ordered pair that satisfies the equation.

- The graph of a function of the form $y = mx + b$, where m and b are constants, is a straight line.

- The graph of a constant function $y = b$, where b is a constant, is a horizontal line.

- The graph of the absolute value function $y = |x|$ is a $\vee$ shape.

- The graph of the squaring function $y = x^2$ is a parabola. More generally, the graph of every function of the form $y = ax^2 + bx + c$, where a, b, and c are real numbers ($a \neq 0$), is a parabola.

- The graph of the equation $x^2 + y^2 = r^2$ is a circle with center at the origin and radius r. An equation of a circle does not define a function.

- When looking at a graph, one can read the domain of a function by considering the variation of the graph in a horizontal direction, while the range is read by considering the variation in the vertical direction.

- *Vertical line test:* Imagine a vertical line sweeping across the graph. If the vertical line at any position intersects the graph in more than one point, the graph is not the graph of a function.

Additional comments

- Be sure to label both axes with appropriate numbers and units. A graph must be able to "stand" by itself.

- It is not necessary to scale the two axes in the same manner. In fact, different scales are frequently desirable. Different scales can have a dramatic effect on the apparent behavior of a relationship.

- Although the Cartesian (or rectangular) coordinate system is most common, there are other systems that can be used to obtain a picture of a relationship.

Detailed solutions to selected exercises

Exercise 10   Find the area of a circle whose diameter extends from $(-1,3)$ to $(3,1)$.

Solution   First, use the distance formula to find the length of the diameter.

$$d = \sqrt{(x_2 - x_1)^2 + (y_2 - y_1)^2} = \sqrt{(3 - (-1))^2 + (1 - 3)^2}$$
$$= \sqrt{16 + 4} = \sqrt{20} = \sqrt{4}\sqrt{5} = 2\sqrt{5}$$

If the diameter is $2\sqrt{5}$, then the radius is $\sqrt{5}$, so

$$\text{Area} = \pi r^2 = \pi(\sqrt{5})^2 = 5\pi.$$

Now do **A**.

Exercise 22   Graph $y = x^2 - 2x + 1$. What is the range of the function?

Solution   Since the equation is of the form $y = ax^2 + bx + c$ (with $a \neq 0$), we know the graph is a parabola. By picking five convenient values of x (say, -2, -1, 0, 1, and 2) and substituting them into $y = x^2 - 2x + 1$, we generate the ordered pairs $(-2,9)$, $(-1,4)$, $(0,1)$, $(1,0)$, and $(2,1)$. By drawing

3.2

A. Find the area of a circle whose diameter extends from $(-2,4)$ to $(3,-1)$.

B. Graph $y = x^2 + 4x + 4$. What is the range of the function?

C. Graph $y = \begin{cases} x \text{ if } x \leq 0 \\ -x \text{ if } x > 0. \end{cases}$
What is the range of the function?

a parabola that passes through these points, we obtain the graph in Figure 3.2:22. The minimum y value is 0 and the graph extends indefinitely in the positive direction, so the range is $[0,\infty)$.

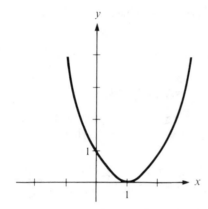

Figure 3.2:22

Now do **B.**

Exercise 44   See text for question.

Solution   If $x \geq 1$, $h(x) = x$ which graphs as a line with ordered pairs $(1,1)$, $(2,2)$, and so on. If $x < -1$, $h(x) = 1$ which is a constant function whose graph is a horizontal line with such ordered pairs as $(-2,1)$, $(-3,1)$, and so on. The graph is given in Figure 3.2:44. The domain of the function is $(-\infty,-1) \cup [1,\infty)$. Since $h(x)$ is greater than or equal to 1, the range is $[1,\infty)$.

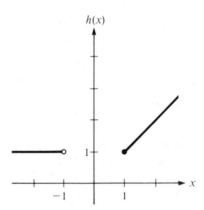

Figure 3.2:44

Now do **C.**

Exercise 50   See text for question.

Solution

a. The x values in the figure start at 0 and keep going to the right. Thus, the domain is $[0,\infty)$.

b. The minimum y value is $-a$ and the graph extends indefinitely in the positive y direction, so there is no maximum value. Thus, the range is $[-a,\infty)$.

c. To determine $f(a)$ and $f(c)$, we want the y values when $x = a$ and $x = c$. The ordered pairs (a,a) and $(c,-a)$ tell us $f(a) = a$ and $f(c) = -a$.

d. From the ordered pairs $(0,a)$, (a,a), and (e,a), we know $y = a$ when $x = 0$, a, or e.

e. $y = 0$ when $x = b$ or d.

f. The y values are less than 0 when the graph is below the x-axis. From the figure we read $f(x) < 0$ when $b < x < d$.

g. The y values are greater than 0 when the graph is above the x-axis. Thus, $f(x) > 0$ when $a < x < b$ or $x > d$.

h. If y is between $-a$ and a, then $|f(x)| < a$. With the exception of the point $(c,-a)$, we read $|y| < a$ when x is between a and e. Thus, $|f(x)| < a$ when $a < x < c$ or $c < x < e$.

Now do **D.**

Exercise 58 See text for question.

Solution The y-axis is a vertical line that intersects the graph in more than one point. Thus by the vertical line test, the graph is not the graph of a function.
Now do **E.**

3.3 Graphing techniques

Objectives checklist

Can you:

a. Classify a function as even, odd, or neither of these?

b. Complete the graph of an even or odd function given the graph for $x \geq 0$?

c. Use the graph of $y = f(x)$ to graph $y = f(x) + c$, $y = f(x) - c$, $y = f(x + c)$, $y = f(x - c)$, $y = -f(x)$, and $y = cf(x)$?

Key terms

Even function   Odd function
Symmetry about the y-axis   Symmetry about the origin

Key rules and formulas

- When x values that are opposite in sign share the same y value (that is, $f(-x) = f(x)$), then f is called an even function and its graph is symmetric with respect to the y-axis.

- When x values that are opposite in sign produce y values that are opposite in sign (that is, $f(-x) = -f(x)$), then f is called an odd function and its graph is symmetric with respect to the origin. In the case of origin symmetry we reflect the curve for $x \geq 0$ about both the x- and y-axis (in either order) to obtain the graph for $x < 0$.

D. Consider the graph of $y = f(x)$ given here.

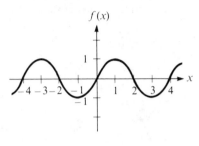

a. What is the domain of the function?
b. What is the range of the function?
c. Determine $f(0)$ and $f(3)$.
d. Solve $f(x) = -1$ on $[0,4]$.
e. Solve $f(x) = 0$ on $[-4,4]$.
f. Solve $f(x) > 0$ on $[0,4]$.
g. For what value(s) of x on $[-4,4]$ is $f(x) < 0$?
h. For what value(s) of x on $[-1,1]$ is $|f(x)| < 1$?

E. Does this graph represent a function?

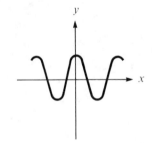

46

3.3

A. Classify the function shown as even, odd, or neither of these.

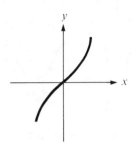

B. Classify the function $y = x^3 - 1$ as even, odd, or neither or these.

- *Vertical and horizontal shifts:* Let c be a positive constant.

  1. The graph of $y = f(x) + c$ is the graph of f raised c units.
  2. The graph of $y = f(x) - c$ is the graph of f lowered c units.
  3. The graph of $y = f(x + c)$ is the graph of f shifted c units to the left.
  4. The graph of $y = f(x - c)$ is the graph of f shifted c units to the right.

- *Reflecting:* The graph of $y = -f(x)$ is the graph of f reflected about the x-axis.

- *Stretching:* If $c > 1$, the graph of $y = cf(x)$ is the graph of f stretched by a factor of c.

- *Shrinking:* If $0 < c < 1$, the graph of $y = cf(x)$ is the graph of f flattened out by a factor of c.

Detailed solutions to selected exercises

Exercise 8   See text for question.

Solution   If we reflect the graph for $x \geq 0$ about both the x-axis and the y-axis (in either order), we obtain the graph for $x < 0$. Therefore, the function is an odd function (and is symmetric with respect to the origin). Now do **A.**

Exercise 12   See text for question.

Solution

a. If f is an even function, then we reflect the graph for $x \geq 0$ about the y-axis to obtain the graph in Figure 3.3:12a.

b. If f is an odd function, then we reflect the graph for $x \geq 0$ about both the x-axis and the y-axis to obtain the graph in Figure 3.3:12b.

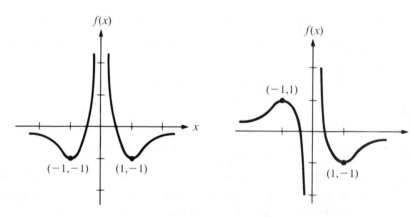

Figure 3.3:12a   **Figure 3.3:12b**

Exercise 22   Classify the function $y = 1/(x^3 + 1)$ as even, odd, or neither of these.

Solution $f(-x) = \dfrac{1}{(-x)^3 + 1} = \dfrac{1}{-x^3 + 1}$. Since $f(-x)$ equals neither $f(x) = \dfrac{1}{x^3 + 1}$ nor $-f(x) = \dfrac{1}{-(x^3 + 1)}$, the function is neither even nor odd.

Now do **B.**

Exercise 30   Use the graph of $y = |x|$ to graph $y = 1 - |x + 1|$.

Solution   The graph of $y = |x|$ is shaped like a $\vee$. The constant 1 is added inside the absolute value so we move our basic shape 1 unit to the left. The constant 1 is also added outside the absolute value (*note:* $1 - |x + 1| = -|x + 1| + 1$), so the $\vee$ moves up 1 unit. Finally, the negative factor in the absolute value term causes a reflection of the graph about the x-axis. The resulting graph is shown in Figure 3.3:30.

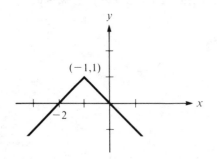

Figure 3.3:30

Figure 3.3:44

Now do **C.**

Exercise 44   See text for question.

Solution   The graph of $y = -f(x)$ is the graph of $y = f(x)$ reflected about the x-axis. The resulting graph is shown in Figure 3.3:44.
Now do **D.**

C. Use the graph of $y = |x|$ to graph $y = -|x + 2|$.

D. Use the graph of $y = f(x)$ shown to graph $y = -f(x)$.

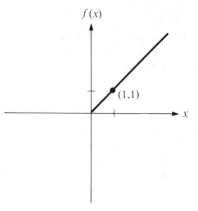

3.4 Operations with functions

Objectives checklist

Can you:

a.   Add, subtract, multiply, and divide two functions and determine the domain in each case?

b.   Find the composite functions of two or more functions and determine the domain in each case?

c.   Rewrite a function h as the composition of simpler functions f and g so that $h(x) = (f \circ g)(x)$?

Key term

Composition

Key rules and formulas

*   For two functions f and g, the functions $f + g, f - g, f \cdot g, f/g$ are defined as follows:

$$(f + g)(x) = f(x) + g(x)$$
$$(f - g)(x) = f(x) - g(x)$$
$$(f \cdot g)(x) = f(x) \cdot g(x)$$
$$\left(\frac{f}{g}\right)(x) = \frac{f(x)}{g(x)} \qquad g(x) \neq 0.$$

3.4

A. If $f(x) = x^4$ and $g(x) = \sqrt[4]{x}$, find $(f + g)(x)$, $(f - g)(x)$, $(f \cdot g)(x)$, $(f/g)(x)$, $(f \circ g)(x)$, and $(g \circ f)(x)$. Indicate the domain of each function.

B. If $h(x) = \sqrt[4]{x^3 + 9}$, find simpler functions f and g so that $h(x) = (f \circ g)(x)$.

C. If $f(x) = x^3$, $g(x) = x + 3$, and $h(x) = 3x$, find $(f \circ g \circ h)(x)$.

The domain of the resulting functions is the intersection of the domains of f and g. The domain of the quotient function excludes any x for which $g(x) = 0$.

- The composite functions of functions f and g are defined as follows:

$$(g \circ f)(x) = g[f(x)]$$
$$(f \circ g)(x) = f[g(x)].$$

The domain of $g \circ f$ is the set of all values in the domain of f for which $g \circ f$ is defined.

Detailed solutions to selected exercises

Exercise 8   See text for question.

Solution   Using the definitions, we have

$$(f + g)(x) = f(x) + g(x) = \sqrt{x} + (-x^2) = \sqrt{x} - x^2$$
$$(f - g)(x) = f(x) - g(x) = \sqrt{x} - (-x^2) = \sqrt{x} + x^2$$
$$(f \cdot g)(x) = f(x) \cdot g(x) = \sqrt{x} \cdot (-x^2) = -x^{5/2}$$
$$(f/g)(x) = f(x)/g(x) = \sqrt{x}/(-x^2) = -1/x^{3/2}$$
$$(f \circ g)(x) = f[g(x)] = f[-x^2] = \sqrt{-x^2}$$
$$(g \circ f)(x) = g[f(x)] = g[\sqrt{x}] = -(\sqrt{x})^2 = -x.$$

The domain of f is $[0,\infty)$ and the domain of g is $(-\infty,\infty)$. The domain of $f + g$, $f - g$, and $f \cdot g$ is the intersection of the domains, which is $[0,\infty)$. In the quotient function, $x \neq 0$, so the domain of f/g is $(0,\infty)$. The domain of $f \circ g$ is $\{0\}$, since zero is the only number in the domain of g for which $f \circ g$ is defined. Finally, the domain of $g \circ f$ is $[0,\infty)$, since the domain of f is limited to nonnegative numbers while $g \circ f$ is defined for all real numbers.
Now do **A.**

Exercise 18   If $h(x) = \sqrt{1 - x^2}$, find simpler functions f and g so that $h(x) = (f \circ g)(x)$.

Solution   The square root and the expression under the square root are the key components. So, if we let $g(x) = 1 - x^2$ and then substitute this function into the square root function $f(x) = \sqrt{x}$, we have

$$(f \circ g)(x) = f[g(x)] = f(1 - x^2) = \sqrt{1 - x^2} = h(x).$$

Thus, $g(x) = 1 - x^2$ and $f(x) = \sqrt{x}$.
Now do **B.**

Exercise 26   If $f(x) = \sqrt{x}$, $g(x) = 1/x$, and $h(x) = 1 - x$, find $(f \circ g \circ h)(x)$.

Solution   Start with function h and use $1 - x$ as a replacement for x in function g, so $(g \circ h)(x) = 1/(1 - x)$. Now use this expression as a replacement for x in function f. Thus, $(f \circ g \circ h)(x) = \sqrt{1/(1 - x)}$.
Now do **C.**

3.5 Inverse functions

Objectives checklist
Can you:

a. Find the inverse of a function?

b. Determine if the inverse of a function is also a function?

c. Find the domain and range of a function and its inverse function?

d. Graph the inverse of a function?

e. Determine whether two functions are inverses of each other?

3.5

A. Find the domain and range of the inverse of function f if $f = \{(3,-1), (4,2), (5,-5)\}$.

Key terms

Inverse functions   One-to-one function

Key rules and formulas

- If (a,b) is an element in a function, then (b,a) is an element in the inverse function. Consequently, the domain of a function f is the range of its inverse function, and the range of f is the domain of its inverse.

- The special symbol f^{-1} is used to denote the inverse function of f.

B. Is the inverse of $\{(1,2), (-1,0), (-2,12)\}$ a function?

- A function is one-to-one when each x value in the domain is assigned a different y value so that no two ordered pairs have the same second component. If f is one-to-one, then the inverse is a function. If f is not one-to-one, then the inverse is not a function.

- *Definition of inverse functions:* Two functions f and g are said to be inverses of each other provided

$$(f \circ g)(x) = f[g(x)] = x \text{ for all } x \text{ in the domain of } g$$

and

$$(g \circ f)(x) = g[f(x)] = x \text{ for all } x \text{ in the domain of } f.$$

- To find the inverse of a function:

  1. Start with a one-to-one function $y = f(x)$ and interchange x and y in this equation.
  2. Solve the resulting equation for y and then replace y by $f^{-1}(x)$.
  3. Define the domain of f^{-1} to be equal to the range of f.

- *Horizontal line test:* Imagine a horizontal line sweeping down the graph of a function. If the horizontal line at any position intersects the graph in more than one point, the function is not one-to-one and its inverse is not a function.

- The graphs of inverse functions are related in that each one is the reflection of the other across the line $y = x$.

Additional comments

- It may be easy to show that a function is one-to-one so that an inverse exists, but finding an equation for the inverse may be difficult or impossible because of the algebra involved.

Detailed solutions to selected exercises

Exercise 4   See text for question.

Solution   By reversing assignments, $f^{-1} = \{(1,-7), (11,-4), (13,1)\}$.
Domain of f = Range of f^{-1} = $\{-7,-4,1\}$;
Range of f = Domain of f^{-1} = $\{1,11,13\}$.
Now do **A.**

Exercise 10   Is the inverse of $\{(-1,1), (0,0), (1,1)\}$ a function?

Solution   No. Since 1 appears as the second component in two ordered pairs, the function is not one-to-one and the inverse is not a function.
Now do **B.**

50

C. Is the inverse of the function whose graph is shown here also a function?

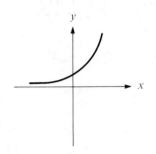

D. Use the graph of $y = f(x)$ shown here to graph $y = f^{-1}(x)$.

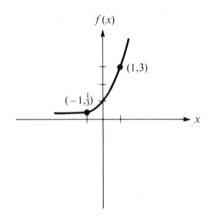

E. Find the inverse function for $y = \sqrt{x + 3}$ and graph both the function and its inverse on the same coordinate system.

Exercise 16   See text for question.

Solution   The inverse of the function graphed is also a function because no horizontal line intersects the graph at more than one point.
Now do **C**.

Exercise 20   See text for question.

Solution   To graph $y = f^{-1}(x)$, we reflect the graph of $y = f(x)$ about the line $y = x$. Note the ordered pairs $(0,1)$, $(\pi/2,0)$, and $(\pi,-1)$ from f become $(1,0)$, $(0,\pi/2)$, and $(-1,\pi)$ in f^{-1}. Both $y = f(x)$ and $y = f^{-1}(x)$ are graphed in Figure 3.5:20.

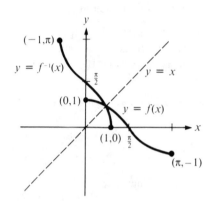

Figure 3.5:20

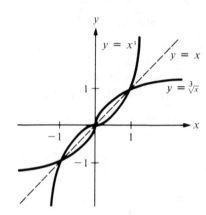

Figure 3.5:26

Now do **D**.

Exercise 26   See text for question.

Solution   The inverse of $y = x^3$ is $x = y^3$ which, when solved for y, becomes $y = \sqrt[3]{x}$. Figure 3.5:26 shows the graph of both $y = x^3$ and $y = \sqrt[3]{x}$.
Now do **E**.

Exercise 46   Verify that $f(x) = 1/(x + 2)$ and $g(x) = (1/x) - 2$ are inverses of each other.

Solution   By the definition of inverse functions, we first show that $f[g(x)] = x$ for all real numbers except 0 (which is the domain of g).

$$f[g(x)] = \frac{1}{[(1/x) - 2] + 2} = \frac{1}{1/x} = x$$

Next, we verify that $g[f(x)] = x$ for all real numbers except -2 (which is the domain of f).

$$g[f(x)] = \frac{1}{1/(x + 2)} - 2 = x + 2 - 2 = x$$

Thus, f and g are inverses of each other.
Now do **F**.

Exercise 52 See text for question.

Solution The inverse of $y = \frac{5}{9}(x - 32)$ is $x = \frac{5}{9}(y - 32)$. Solving for y, we determine the inverse function is $y = \frac{9}{5}x + 32$. This function converts degrees Celsius to degrees Fahrenheit.
Now do **G**.

3.6 Slope and rate of change

Objectives checklist

Can you:

a. Determine the rate of change in a linear relationship by finding the slope?

b. Interpret the meaning of slope in a graph and in applied problems?

c. Find the average rate of change of a function for a given interval?

Key terms

Linear function   Secant line
Slope of a line   Average rate of change of a function

Key rules and formulas

- *Definition of slope:* If (x_1, y_1) and (x_2, y_2) are any two points on a nonvertical line, then

$$\text{slope} = m = \frac{\Delta y}{\Delta x} = \frac{y_2 - y_1}{x_2 - x_1}.$$

(*Note:* The symbol Δ is the Greek capital letter delta. The symbol is used to indicate a change.)

- The slope of a line may be interpreted as a measure of the inclination or steepness of a line. We often refer to Δy as the "rise" and Δx as the "run" so that

$$\text{slope} = \frac{\text{rise}}{\text{run}}.$$

- The slope of every horizontal line is zero.

- The slope of every vertical line is undefined.

- A positive slope indicates that y is increasing as x increases.

- A negative slope indicates that y is decreasing as x increases.

- A zero slope indicates that y remains the same as x increases. In this case the line is horizontal and the function is a constant function.

- The average rate of change of a function f as x varies from x_1 to x_2 is

$$\frac{\Delta y}{\Delta x} = \frac{y_2 - y_1}{x_2 - x_1} = \frac{f(x_2) - f(x_1)}{x_2 - x_1}.$$

- The average rate of change of a function f as x varies from x to $x + h$ is

$$\frac{\Delta y}{\Delta x} = \frac{f(x + h) - f(x)}{h}, h \neq 0.$$

F. Verify that $f(x) = 1/(x - 5)$ and $g(x) = (1/x) + 5$ are inverses of each other.

G. The function
$y = \frac{9}{5}(x - 273.16) + 32$
converts degrees Kelvin (x) to degrees Fahrenheit (y). Find the inverse of this function. What formula does the inverse function represent?

52

3.6

A. Find the slope of the line determined by $(-2,-3)$ and $(-4,-7)$.

B. The college bookstore sells a textbook that costs $10 for $13.50 and a textbook that costs $12 for $15.90. Find the rate of change between cost and selling price if the relationship is linear by calculating slope. Interpret the meaning of the slope.

Detailed solutions to selected exercises

Exercise 6   Find the slope of the line determined by $(-5,-1)$ and $(-2,-4)$.

Solution   If we label $(-5,-1)$ as point 1, then $x_1 = -5$, $y_1 = -1$, $x_2 = -2$, and $y_2 = -4$. Then

$$m = \frac{y_2 - y_1}{x_2 - x_1} = \frac{(-4) - (-1)}{(-2) - (-5)} = \frac{-3}{3} = -1.$$

A slope of -1 means that as x increases 1 unit, y decreases 1 unit.
Now do **A.**

Exercise 16a   See text for question.

Solution   Consider line a in the figure in the text. Note that the points $(-3,3)$ and $(2,2)$ are on the line so that as x increases 5 units (from -3 to 2), y decreases 1 unit (from 3 to 2). Thus, $m = \Delta y/\Delta x = -\frac{1}{5}$.

Exercise 18   On a piece of graph paper draw lines through the point $(-2,-1)$ with the following slopes:
(a) 1   (b) -4 (c) $-\frac{1}{3}$ (d) $\frac{1}{3}$ (e) $-\frac{3}{2}$.

Solution

a. Since $m = 1$, as x increases 1, y increases 1. Starting at $(-2,-1)$ and increasing both x and y by 1 gives $(-1,0)$ as a second point on the line. Now draw a line through these two points as in Figure 3.6:18.
b. Since $m = -4$, as x increases 1, y decreases 4. Starting at $(-2,-1)$, we go 1 unit to the right and 4 units down, so $(-1,-5)$ is a second point on the line (see Figure 3.6:18).
c.–e. Slopes of $-\frac{1}{3}$, $\frac{1}{3}$, and $-\frac{3}{2}$ produce second points with coordinates $(1,-2)$, $(1,0)$, and $(0,-4)$, respectively. The resulting lines are shown in Figure 3.6:18.

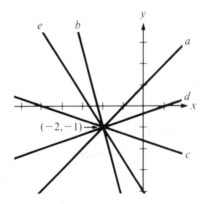

Figure 3.6:18

Exercise 20   See text for question.

Solution   We are given $C_1 = 0$, $F_1 = 32$ and $C_2 = 100$, $F_2 = 212$. Then

$$m = \frac{\Delta C}{\Delta F} = \frac{100 - 0}{212 - 32} = \frac{5}{9} \quad \text{or} \quad m = \frac{\Delta F}{\Delta C} = \frac{212 - 32}{100 - 0} = \frac{9}{5}.$$

Thus, a $9°$ increase in Fahrenheit temperature is a $5°$ increase in Celsius temperature.
Now do **B.**

Exercise 28   Find the average rate of change of $y = \sqrt{x}$ from $x_1 = 4$ to $x_2 = 9$.

Solution   Substituting in the formula for average rate of change gives

$$\frac{\Delta y}{\Delta x} = \frac{f(x_2) - f(x_1)}{x_2 - x_1} = \frac{\sqrt{9} - \sqrt{4}}{9 - 4} = \frac{3 - 2}{5} = \frac{1}{5}.$$

Now do **C**.

Exercise 38   See text for question.

Solution   Since $f(x) = \sqrt{x}$, we have $f(x + h) = \sqrt{x + h}$, so

$$\frac{f(x + h) - f(x)}{h} = \frac{\sqrt{x + h} - \sqrt{x}}{h}.$$

Using the hint to rationalize the numerator gives

$$\frac{\sqrt{x + h} - \sqrt{x}}{h} \cdot \frac{\sqrt{x + h} + \sqrt{x}}{\sqrt{x + h} + \sqrt{x}} = \frac{(x + h) - x}{h(\sqrt{x + h} + \sqrt{x})}$$

$$= \frac{h}{h(\sqrt{x + h} + \sqrt{x})} = \frac{1}{\sqrt{x + h} + \sqrt{x}}.$$

Now do **D**.

Exercise 42   See text for question.

Solution   We modify the standard formula by using t in place of x and fixing t specifically at 1. So

$$\frac{f(x + h) - f(x)}{h} \quad \text{becomes} \quad \frac{f(t + h) - f(t)}{h},$$

$$\text{which becomes} \quad \frac{f(1 + h) - f(1)}{h}.$$

Now since $f(t) = 16t^2$, the average velocity ($\Delta d / \Delta t$) from 1 to $1 + h$ is

$$\frac{16(1 + h)^2 - 16(1)^2}{h} = \frac{16 + 32h + 16h^2 - 16}{h}$$

$$= \frac{32h + 16h^2}{h} = 32 + 16h.$$

When $h = 0.1$, 0.01, and 0.001 seconds, the average velocities are 33.6 ft/sec, 32.16 ft/sec, and 32.016 ft/sec, respectively. At $t = 1$, we say the velocity is 32 ft/sec.

Now do **E**.

C. Find the average rate of change of $y = \sqrt[3]{x}$ from $x = 8$ to $x = 27$.

D. If $f(x) = \sqrt{x + 2}$, compute $\dfrac{f(x + h) - f(x)}{h}$ with $h \neq 0$ to determine the average rate of change of f from x to $x + h$.

E. Suppose the distance (d) in meters traveled by an object in t seconds is given by the formula $d = 4.9t^2$. What is the average velocity of the object in the interval from $t_1 = 2$ to $t_2 = 2 + h$? What is the average velocity if $h = 0.1$? 0.01? 0.001? Intuitively, what do you think we say the velocity of the object is at $t = 2$?

3.7 Variation

Objectives checklist

Can you:

a. Write an equation that expresses the functional relationship between variables in the language of variation?

b. Solve applied problems involving variation?

Key terms

Direct variation   Variation constant
Inverse variation

3.7

A. If y varies directly as x and inversely as z, and $y = 9$ when $x = 3$ and $z = 3$, find y when $x = 4$ and $z = 7$.

B. The intensity of light on a plane surface varies inversely as the square of the distance from the source of the light. If we halve the distance from the source to the plane, what happens to the intensity?

C. The safe load of a beam varies directly as the width and the square of the height and inversely as the distance between the supports at each end. If the width is tripled, the height is halved, and the distance between the supports remains the same, what is the effect on the safe load?

Key rules and formulas

- The statement "y varies directly as x" means there is some positive number k (variation constant) such that $y = kx$.

- The statement "y varies inversely as x" means there is some positive number k such that $xy = k$ or $y = k/x$.

Additional comments

- The concept of variation extends to include direct and inverse variation of variables raised to specified powers and relationships that involve more than two variables.

- The variation constant k is often called the constant of proportionality and the expression "varies directly as" is often replaced by "is proportional to."

Detailed solutions to selected exercises

Exercise 10 If y varies directly as x and inversely as z, and $y = 10$ when $x = 4$ and $z = 3$, find y when $x = 7$ and $z = 15$.

Solution Since y varies directly as x and inversely as z, we have $y = kx/z$. To find k, replace y by 10, x by 4, and z by 3.

$$10 = \frac{k \cdot 4}{3} \text{ so } k = \frac{15}{2}.$$

Then $y = \dfrac{15}{2} \cdot \dfrac{x}{z}$ so, when $x = 7$ and $z = 15$, $y = \dfrac{15}{2} \cdot \dfrac{7}{15} = \dfrac{7}{2}$.

Now do **A.**

Exercise 24 See text for question.

Solution The intensity I varies inversely as the square of the distance, so $I = k/d^2$. If the distance is doubled, we have $I = k/(2d)^2 = k/(4d^2)$, which means the intensity is divided by 4.
Now do **B.**

Exercise 28 See text for question.

Solution If S represents the safe load, w and h the width and height of the beam, and d the distance between supports, then the given variations tell us $S = kwh^2/d$. If the width and height are doubled and the distance remains the same, then

$$S = \frac{k(2w)(2h)^2}{d} = \frac{8kwh^2}{d}.$$

Therefore, the safe load is multiplied by 8 under the conditions given.
Now do **C.**

Sample test questions: Chapter 3

1. Complete the statement.

   a. The domain of $y = 1/\sqrt{x^2 - 1}$ is _____ .

   b. The range of $y = x^2 - 3$ is _____ .

   c. If $g(x) = 2 - 3x$, then $g(x) = 4$ when x equals _____ .

   d. The function expressing the radius (r) of a circle as a function of its circumference (C) is _____ .

e. If $g(x) = 2x$ and $f(x) = \dfrac{1}{2x}$ then $(g \circ f)(x)$ equals _____ .

f. If $f(x) = 2x^2 - 5x$, then $f(2)$ equals _____ .

g. One ordered pair in f if $f(x) = 1$ is _____ .

h. The distance between $(-4,7)$ and $(6,3)$ is _____ .

i. If $f(x) = (x/2) + 1$, then $f^{-1}(x)$ equals _____ .

j. If y varies inversely as x, and $y = 7$ when $x = 10$, then the y value when $x = 20$ is _____ .

2. Graph the given function.

   a. $y = (x + 2)^2$ b. $f(x) = 1 - x^2$

   c. $y = |x + 3|$ d. $y = 2x - 4$

   e. $y = x^3 + 2$ f. $y = 2$

3. Answer the following questions using the accompanying figure.

   a. What is the domain?

   b. What is the range?

   c. Find $f(0)$.

   d. Solve $f(x) = 0$ on $[-2,2]$.

   e. Solve $f(x) < 0$ on $[-2,2]$.

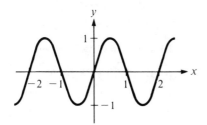

4. If $f(x) = \begin{cases} x \text{ if } -1 \le x \le 1 \\ 1 \text{ if } x > 1 \end{cases}$, then

   a. Find $f(0)$ b. What is the domain?

   c. What is the range?

5. Compute $[f(x + h) - f(x)]/h$ with $h \ne 0$ if $f(x) = 1 - 2x^2$.

6. Find the area of the right triangle with vertices at $(0,0)$, $(-4,1)$, and $(-5,-3)$.

7. a. The volume of a sphere varies directly as the cube of its radius. If the radius is doubled, then by what factor does the volume change?

   b. Express the federal income tax (t) for single persons as a function of their taxable income (i) if their taxable income is between $4,000 and $6,000 (inclusive) and the tax rate is $690 plus 21 percent of the excess over $4,000. What is the domain of the function?

8. Select the choice that completes the statement or answers the question.

a. If $f(x) = 2x - 4$ and $g(x) = x^2 + 1$, then $(f - g)(-3)$ equals

    (a) -8 (b) 0   (c) 6   (d) -20 (e) -18

b. If $f(x) = x + 2$, does $f(x + 2) = f(x) + f(2)$?

    (a) yes   (b) no

c. The graph of $y = x^3$ is symmetric with respect to

    (a) the x-axis   (b) the y-axis

    (c) the origin   (d) none of these

d. The graph of an even function is symmetric with respect to

    (a) the x-axis   (b) the y-axis

    (c) the origin   (d) none of these

e. If $f(x) = 2x + 2$, then $f^{-1}(2)$ equals

    (a) 6   (b) -6 (c) 2   (d) -2 (e) 0

f. The graph of $y = f(x + 2)$ is the graph of $y = f(x)$ shifted 2 units

    (a) up   (b) down   (c) left   (d) right

g. Which number is not in the domain of $y = (x - 1)/(x - 3)$?

    (a) 3   (b) -3 (c) 0   (d) 1   (e) -1

h. If $(a,0)$ is a solution of $y = -2x + 3$, then a equals

    (a) $-\frac{2}{3}$ (b) -2 (c) 3   (d) $\frac{2}{3}$ (e) $\frac{3}{2}$

i. If y varies directly as x, then when x is doubled y is

    (a) increased by 2   (b) decreased by 2

    (c) multiplied by 2   (d) divided by 2

Chapter 4
Polynomial and Rational Functions

A. Is $f(x) = 3x^2 + 2x^{1/2} + 4x - 6$ a polynomial function?

4.1 Linear functions

Objectives checklist

Can you:

a. Determine if a function is a polynomial function?

b. Write an equation for a line when given either the slope and the y-intercept, or the slope and a point on the line, or two points on the line?

c. Find the slope and y-intercept of the line defined by a given equation?

d. Find the equation of the linear function f given two ordered pairs in the function?

e. Graph a line whose slope and y-intercept are given?

f. Find an equation for the line passing through a given point that is parallel to a given line?

g. Find an equation for the line passing through a given point that is perpendicular to a given line?

h. Use the concept of slope to determine if three points are the vertices of a right triangle or if three points lie on a line?

i. Solve applied problems involving linear functions?

Key terms

Polynomial function
Linear function
y-intercept

Slope-intercept equation of a line
Point-slope equation of a line
General equation of a line

Key rules and formulas

* *Point-slope form:* An equation of the line through (x_1, y_1) with slope m is

$$y - y_1 = m(x - x_1).$$

* *Slope-intercept form:* The graph of the equation

$$y = mx + b$$

is a line with slope m and y-intercept $(0, b)$.

* *Parallel lines:* Two nonvertical lines are parallel if and only if their slopes are equal.

* *Perpendicular lines:* Two nonvertical lines are perpendicular if and only if the product of their slopes is -1.

Detailed solutions to selected exercises

Exercise 6   Is $f(x) = x^{-3} + 2x^2 - x + 3$ a polynomial function?

Solution   In a polynomial function the exponent above x in each term must be a nonnegative integer. Therefore, the given function is not a polynomial function because of the term x^{-3}.
Now do **A.**

58

B. Find an equation for the line through $(3,-5)$ with slope $-\frac{2}{5}$.

C. Find an equation for the line through $(8,-6)$ and $(2,-6)$.

D. If $3x - 4y = -8$, find the slope and y-intercept.

E. Graph the line with slope 2 and y-intercept $(0,-4)$ and find an equation for the line.

Exercise 14 Find an equation for the line through $(-5,2)$ with slope $-\frac{3}{4}$.

Solution When the slope and one point of the line are given, the point-slope equation $y - y_1 = m(x - x_1)$ is a natural choice. Substituting

$x_1 = -5$, $y_1 = 2$, and $m = -\frac{3}{4}$ in the equation gives

$$y - 2 = -\frac{3}{4}[x - (-5)] \text{ so } y - 2 = -\frac{3}{4}x - \frac{15}{4} \text{ so } y = -\frac{3}{4}x - \frac{7}{4}.$$

Now do **B**.

Exercise 20 Find an equation for the line through $(2,-3)$ and $(3,-3)$.

Solution Since the y-coordinate is the same for both points, the line is horizontal and the y value is fixed at -3. Thus, the equation for the line is $y = -3$. You should note that when $m = 0$, $y = mx + b$ simplifies to $y = b$. Now do **C**.

Exercise 30 If $2x - 7y = -4$, find the slope and the y-intercept.

Solution First, transform the equation to the form $y = mx + b$.

$$2x - 7y = -4 \text{ so } -7y = -2x - 4 \text{ so } y = \frac{2}{7}x + \frac{4}{7}$$

Thus, $m = \frac{2}{7}$ and $b = \frac{4}{7}$, so the slope is $\frac{2}{7}$ and the y-intercept is $(0,\frac{4}{7})$. Now do **D**.

Exercise 34 Graph the line with slope $m = -\frac{3}{4}$ and y-intercept $(0,-2)$ and find an equation for the line.

Solution One point on the line is $(0,-2)$. Since the slope is $-\frac{3}{4}$, by starting at $(0,-2)$ and going 4 units to the right and 3 units down, we obtain a second point on the same line at $(4,-5)$. We join these points with a line to get the graph in Figure 4.1:34.

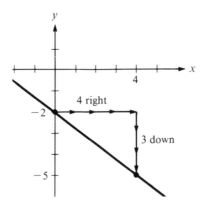

Figure 4.1:34

Substituting $m = -\frac{3}{4}$ and $b = -2$ in the slope-intercept equation $y = mx + b$, we have $y = -\frac{3}{4}x - 2$ as an equation for the line. Now do **E**.

Exercise 38 Find the equation that defines the linear function f in which $f(0) = -1$ and $f(2) = 0$.

Solution If $f(0) = -1$, then when $x = 0$, $y = -1$. Similarly, if $f(2) = 0$, then when $x = 2$, $y = 0$. Substituting these values in the slope formula gives

$$m = \frac{y_2 - y_1}{x_2 - x_1} = \frac{0 - (-1)}{2 - 0} = \frac{1}{2}.$$

To complete the solution, note that $(0,-1)$ is the y-intercept, so from the slope-intercept equation $y = mx + b$ we have $y = \frac{1}{2}x - 1$.
Now do **F**.

Exercise 54 Find an equation for the line through $(-1,-4)$ that is (a) parallel and (b) perpendicular to $4x - y = -8$.

Solution

a. In the form $y = mx + b$ we write $4x - y = -8$ as $y = 4x + 8$, so the slope of the given line is 4. Since the slopes of parallel lines are equal, we want an equation for the line through $(-1,-4)$ with slope 4. By using the point-slope equation $y - y_1 = m(x - x_1)$, we have

$$y - (-4) = 4(x - (-1)) \text{ so } y + 4 = 4x + 4 \text{ so } y = 4x.$$

b. The slope of the given line is 4. Since the slopes of perpendicular lines are negative reciprocals of each other, the perpendicular line has slope $-\frac{1}{4}$. Then from the point-slope equation we have

$$y - (-4) = -\tfrac{1}{4}(x - (-1)) \text{ so } y + 4 = -\tfrac{1}{4}x - \tfrac{1}{4} \text{ so } y = -\tfrac{1}{4}x - \tfrac{17}{4}.$$

Now do **G**.

Exercise 62 See text for question.

Solution

a. The length of the spring is a function of the force applied. If x represents the number of pounds of force applied and y represents the length of the spring, then we are given that $x_1 = 16$ lb, $y_1 = 20$ in. and $x_2 = 36$ lb, $y_2 = 15$ in. By the slope formula, we then have

$$m = \frac{y_2 - y_1}{x_2 - x_1} = \frac{15 - 20}{36 - 16} = \frac{-5}{20} = -\frac{1}{4}.$$

Finally, substituting $x_1 = 16$, $y_1 = 20$, and $m = -\frac{1}{4}$ in the point-slope equation gives

$$y - 20 = -\tfrac{1}{4}(x - 16), \text{ which simplifies to } y = -\tfrac{1}{4}x + 24.$$

b. If $x = 28$ lb, $y = -\frac{1}{4}(28) + 24 = 17$ in.

c. If $y = 10$ in., then $-\frac{1}{4}x + 24 = 10$, so $x = 56$ lb.
Now do **H**.

Exercise 64 Show the points $A(-7,-7)$, $B(-1,1)$, and $C(2,5)$ lie on a straight line by using the concept of slope.

Solution Find the slopes of line segments AB and BC.

$$m_{AB} = \frac{1 - (-7)}{-1 - (-7)} = \frac{8}{6} = \frac{4}{3} \text{ while } m_{BC} = \frac{5 - 1}{2 - (-1)} = \frac{4}{3}.$$

Since the line segments are connected and parallel (same slope), the points lie on a line.
Now do **I**.

F. Find the equation that defines the linear function f in which $f(0) = -3$ and $f(-3) = 6$.

G. Find an equation for the line through $(-3,-8)$ that is (a) parallel and (b) perpendicular to $-3x + y = -12$.

H. A certain fish tank weighs 140 lb when it contains 10 gal of water and 239 lb when it contains 21 gal of water.
  a. Find an equation that defines this relationship if it is linear.
  b. What would the tank weigh if it contained 15 gal of water?
  c. How much did the tank weigh before the water was added?
  d. How much did the water weigh per gallon?

I. Show that the points $A(5,4)$, $B(1,2)$, and $C(-7,-2)$ lie on a straight line by using the concept of slope.

Exercise 68 See text for question.

Solution The slope of $y = x$ is 1. The slope of line segment AB is

$$m = \frac{y_2 - y_1}{x_2 - x_1} = \frac{a - b}{b - a} = -1.$$

Since the product of the slopes is -1, line segment AB is perpendicular to $y = x$. Also, using the points in the figure and the distance formula, we have

$$d_1 = \sqrt{(c - a)^2 + (c - b)^2} \quad \text{and} \quad d_2 = \sqrt{(b - c)^2 + (a - c)^2}.$$

Since $d_1 = d_2$, points A and B are equidistant from L. Thus, the two conditions in the definition are satisfied, so (a,b) and (b,a) are symmetric about $y = x$.

4.2 Quadratic functions

Objectives checklist

Can you:

a. Graph a quadratic function and indicate on the graph the x- and y-intercepts, the axis of symmetry, and the coordinates of the maximum or minimum point?

b. Find the range of a quadratic function?

c. Determine the vertex and axis of symmetry of the graph of a quadratic function by matching the function to the form $f(x) = a(x - h)^2 + k$?

d. Solve applied problems involving quadratic functions?

e. Solve a quadratic inequality by graphing $f(x) = ax^2 + bx + c$ and reading from the graph values of x for which $f(x) < 0$ or $f(x) > 0$?

Key terms

Quadratic function   Vertex
Axis of symmetry

Key rules and formulas

- A function of the form $f(x) = ax^2 + bx + c$ (where a, b, and c are real numbers with $a \neq 0$) is called a quadratic function. The graph of a quadratic function is a parabola. In general:

    1. If a is positive ($a > 0$), the parabola opens upward and the vertex is a minimum point.
    2. If a is negative ($a < 0$), the parabola opens downward and the vertex is a maximum point.

- The equation of the axis of symmetry is

$$x = -b/2a.$$

- The coordinates of the vertex of the parabola are

$$(-b/2a, f(-b/2a)).$$

- The x-intercepts are found by solving the equation $ax^2 + bx + c = 0$. The y-intercept is always $(0,c)$.

- Any quadratic function may be placed in the form

$$f(x) = a(x - h)^2 + k$$

by completing the square. The vertex is the point (h,k) and the axis of symmetry is the vertical line $x = h$.

- To solve a quadratic inequality by graphing, do the following:

    1. Sketch the graph of $f(x) = ax^2 + bx + c$ with emphasis on the x-intercepts and on whether the parabola has a maximum or minimum point. The actual coordinates of the vertex need not be found.
    2. To solve $f(x) < 0$, specify all x values for which the parabola is below the x-axis. To solve $f(x) > 0$, read from the graph all x values for which the parabola is above the x-axis.

Additional comments

- An alternate approach to solving inequalities with quadratic expressions that factor was given in Section 2.5.

Detailed solutions to selected exercises

Exercise 18   See text for question.

Solution

a. To find the x-intercepts, set $y = 0$ and use the quadratic formula.

$$3x^2 - 5x + 1 = 0 \text{ so } x = \frac{-(-5) \pm \sqrt{(-5)^2 - 4(3)(1)}}{2(3)} = \frac{5 \pm \sqrt{13}}{6}$$

x-intercepts: $((5 + \sqrt{13})/6, 0)$ and $((5 - \sqrt{13})/6, 0)$
y-intercepts: $(0,1)$ since $c = 1$

b. Axis of symmetry: $x = \dfrac{-b}{2a} = \dfrac{-(-5)}{2(3)} = \dfrac{5}{6}$

c. Since $a > 0$, we have a minimum point. The x-coordinate of the minimum point is $\frac{5}{6}$ since the minimum point lies on the axis of symmetry. We find the y value of the minimum point by finding the y value when $x = \frac{5}{6}$.

$$\begin{aligned} y = f(x) &= 3x^2 - 5x + 1 \\ f(\tfrac{5}{6}) &= 3(\tfrac{5}{6})^2 - 5(\tfrac{5}{6}) + 1 \\ &= \tfrac{75}{36} - \tfrac{25}{6} + 1 \\ &= -\tfrac{39}{36} = -\tfrac{13}{12} \end{aligned}$$

Thus, the minimum point is $(\frac{5}{6}, -\frac{13}{12})$.

d. We read in Figure 4.2:18 that the range is $[-\frac{13}{12}, \infty)$.

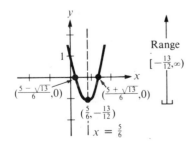

Figure 4.2:18

Now do **A**.

4.2

A. Graph the function $y = 2x^2 - x - 1$ and indicate
   a. the coordinates of the x- and y-intercepts
   b. the equation of the axis of symmetry
   c. the coordinates of the maximum or minimum point
   d. the range of the function

B. Determine the vertex and axis of symmetry of the graph of $f(x) = 2x^2 - 4x + 7$ by matching the function to the form $f(x) = a(x - h)^2 + k$.

Exercise 28   Determine the vertex and axis of symmetry of the graph of $f(x) = 2x^2 + 4x - 5$ by matching the function to the form $f(x) = a(x - h)^2 + k$.

Solution   We first factor out 2 from the x terms and then complete the square as follows:

$$f(x) = 2(x^2 + 2x) - 5 = 2(x^2 + 2x + 1) - 7 = 2(x + 1)^2 - 7.$$

We match this equation to the form $f(x) = a(x - h)^2 + k$, which gives $h = -1$ and $k = -7$. Thus, the vertex is $(-1, -7)$ and the axis of symmetry is $x = -1$.
Now do **B**.

C. Find the maximum possible area of a rectangle with a perimeter of 200 ft.

Exercise 44   Find the maximum possible area of a rectangle with a perimeter of 100 ft.

Solution   The area and perimeter formulas for a rectangle are

$$A = \ell w \quad \text{and} \quad P = 2\ell + 2w.$$

Since $P = 100$ ft, we know $100 = 2\ell + 2w$ so $\ell = 50 - w$. Substituting this expression for ℓ in the area formula gives

$$A = (50 - w)w = 50w - w^2.$$

The area is a maximum when

$$w = -b/2a = -50/2(-1) = 25,$$

D. Solve $8 \geq 2x + x^2$.

so the maximum value for the area is

$$A = (50 - 25)25 = 625 \text{ ft}^2.$$

Note the maximum area results from a rectangle that is a square.
Now do **C**.

Exercise 52   Solve $-10 \geq 3x - x^2$.

Solution $-10 \geq 3x - x^2$ in the form $ax^2 + bx + c \geq 0$ is written as $x^2 - 3x - 10 \geq 0$. Therefore we need a rough sketch of the graph of $y = x^2 - 3x - 10$. For the x-intercepts we have

E. Solve $3x^2 + 2x > -2$.

$$0 = (x - 5)(x + 2) \quad \text{so} \quad x = 5 \text{ and } x = -2.$$

Since $a > 0$, the vertex is a minimum point. Now sketch Figure 4.2:52. To solve $f(x) \geq 0$, we specify all x values for which the parabola is on or above the x-axis. Thus, we read the solution set from the graph to be $(-\infty, -2] \cup [5, \infty)$.
Now do **D**.

Exercise 66   Solve $-2x^2 + x > 4$.

Solution   First, rewrite $-2x^2 + x > 4$ as $-2x^2 + x - 4 > 0$. Now find the x-intercepts for $y = -2x^2 + x - 4$.

$$x = \frac{-1 \pm \sqrt{(1)^2 - 4(-2)(-4)}}{2(-2)} = \frac{-1 \pm \sqrt{-31}}{-4} \text{ (not real numbers)}$$

Since there are no x-intercepts and since $a < 0$ indicates the vertex is a maximum point, the graph of $y = -2x^2 + x - 4$ is always below the x-axis as shown in Figure 4.2:66. Thus, there are no values of x for which $-2x^2 + x - 4$ is greater than 0, so the solution set is $\emptyset$.
Now do **E**.

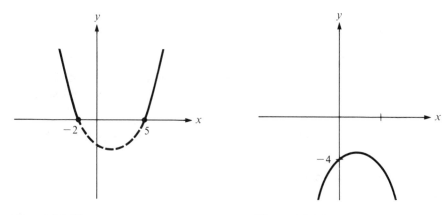

Figure 4.2:52   **Figure 4.2:66**

4.3 Synthetic division and the remainder theorem

Objectives checklist

Can you:

a. Divide one polynomial by another using long division?

b. Divide any polynomial by a polynomial of the form $x - b$ using synthetic division?

c. Find the value of a polynomial function for a specified x by both direct substitution and the remainder theorem?

Key term

Synthetic division

Key rules and formulas

- *Long division:* To divide one polynomial by another, do the following:

  1. Arrange the terms of the dividend and the divisor in descending powers of x. If a term is missing in the dividend, write 0 as its coefficient.
  2. Divide the first term of the dividend by the first term of the divisor to obtain the first term of the quotient.
  3. Multiply the entire divisor by the first term of the quotient and subtract this product from the dividend.
  4. Use the remainder as the new dividend and repeat the above procedure until the remainder is of lower degree than the divisor.

- *Synthetic division:* To divide by $x - b$, use the following shortcut method.

  1. Write the coefficients of the terms in the dividend. Be sure the powers are in descending order and enter 0 as the coefficient of any missing term. Also, write the value of b to the left of the dividend.
  2. Bring down the first dividend entry. Multiply this number by b and place the result under the second coefficient. Add, then multiply this result by b. Place the result under the third coefficient, and so on.
  3. The last number in the bottom row is the remainder. The other numbers in the bottom row are the coefficients of the quotient. The degree of the polynomial in the quotient is always one less than the degree of the polynomial in the dividend.

4.3

A. Divide $3x^3 - x^2 + 6x - 8$ by $x^2 + 2$.

B. Divide by synthetic division $(3x^3 + 2x^2 + 1) \div (x + 2)$.

- *Remainder theorem:* If a polynomial $P(x)$ is divided by $x - b$, the remainder is $P(b)$.

Additional comments

- A summary cannot do justice to the simple procedures in long division and synthetic division. Careful consideration of the example problems should help clarify the steps.

- Synthetic division applies only when we divide by a polynomial of the form $x - b$.

Detailed solutions to selected exercises

Exercise 6 Divide $(2x^3 - 3x^2 + 10x - 5) \div (x^2 + 5)$.

Solution The long division procedure is as follows:

$$
\begin{array}{r}
2x - 3 \\
x^2 + 5\overline{)2x^3 - 3x^2 + 10x - 5} \\
\underline{2x^3 + 10x} \text{(subtract)} \\
-3x^2 -5 \\
\underline{-3x^2 - 15} \text{ (subtract)} \\
10
\end{array}
$$

The answer in the form requested in the text is

$$2x^3 - 3x^2 + 10x - 5 = (x^2 + 5)(2x - 3) + 10.$$

Now do **A.**

Exercise 10 Divide by synthetic division $(3x^4 - x^2 + 7) \div (x + 3)$.

Solution We use 0's as the coefficients of the missing x^3 and x terms in the dividend. In the form $x - b$ the divisor $x + 3$ is $x - (-3)$, so $b = -3$. The synthetic division procedure is as follows:

$$
\begin{array}{r|rrrrr}
-3| & 3 & 0 & -1 & 0 & 7 \\
& & -9 & 27 & -78 & 234 \\
\hline
& 3 & -9 & 26 & -78 & 241 \quad \leftarrow \text{ remainder}
\end{array}
$$

quotient $\rightarrow 3x^3 - 9x^2 + 26x - 78$.

The answer in the form requested in the text is

$$3x^4 - x^2 + 7 = (x + 3)(3x^3 - 9x^2 + 26x - 78) + 241.$$

Now do **B.**

Exercise 18 If $P(x) = 2x^4 - x^3 + 2x - 1$, find $P(\frac{1}{2})$ and $P(-\frac{1}{2})$ by (a) direct substitution and (b) the remainder theorem.

Solution
a. By direct substitution,

$$
\begin{aligned}
P(\tfrac{1}{2}) &= 2(\tfrac{1}{2})^4 - (\tfrac{1}{2})^3 + 2(\tfrac{1}{2}) - 1 \\
&= 2(\tfrac{1}{16}) - (\tfrac{1}{8}) + 1 - 1 = 0 \\
P(-\tfrac{1}{2}) &= 2(-\tfrac{1}{2})^4 - (-\tfrac{1}{2})^3 + 2(-\tfrac{1}{2}) - 1 \\
&= 2(\tfrac{1}{16}) + (\tfrac{1}{8}) - 1 - 1 = -\tfrac{7}{4}.
\end{aligned}
$$

b. By the remainder theorem, we find $P(b)$ by determining the remainder when $P(x)$ is divided by $x - b$. Since this remainder may be obtained by the shortcut method of synthetic division, this approach is simple and often provides us with important information about $P(x)$ as discussed in Sections 4.4 and 4.5.

For $P(\frac{1}{2})$:

$$\frac{1}{2} \rfloor \quad 2 \quad -1 \quad 0 \quad 2 \quad -1$$
$$\phantom{\frac{1}{2} \rfloor \quad 2} \quad 1 \quad 0 \quad 0 \quad 1$$
$$\phantom{\frac{1}{2} \rfloor} \quad 2 \quad 0 \quad 0 \quad 2 \quad 0 \quad \leftarrow \text{remainder}$$

Since the remainder is 0, $P(\frac{1}{2}) = 0$.

For $P(-\frac{1}{2})$:

$$-\frac{1}{2} \rfloor \quad 2 \quad -1 \quad 0 \quad 2 \quad -1$$
$$\phantom{-\frac{1}{2} \rfloor \quad 2} \quad -1 \quad 1 \quad -\frac{1}{2} \quad -\frac{3}{4}$$
$$\phantom{-\frac{1}{2} \rfloor} \quad 2 \quad -2 \quad 1 \quad \frac{3}{2} \quad -\frac{7}{4} \quad \leftarrow \text{remainder}$$

Since the remainder is $-\frac{7}{4}$, $P(-\frac{1}{2}) = -\frac{7}{4}$.

Now do **C**.

C. If $P(x) = 3x^4 - 4x^3 + x^2 - 3$, find $P(\frac{1}{3})$ and $P(-\frac{1}{3})$ by
a. direct substitution
b. the remainder theorem

4.4 Theorems about zeros

Objectives checklist

Can you:

a. Find the zeros of a polynomial function in factored form and state the multiplicity of each zero and the degree of the polynomial?

b. Write a polynomial function (in factored form) given its degree and zeros?

c. Graph a polynomial function based on information obtained from the intercepts?

d. Write a polynomial function (in factored form) with rational coefficients of the lowest possible degree with given zeros?

e. Find all the zeros in a polynomial function when given a sufficient number of its zeros so the remaining zeros can be determined by the theorems in this section?

Key terms

Zero of a function   Multiplicity of a zero

Key rules and formulas

- *Factor theorem:* If b is a zero of the polynomial function $y = P(x)$, then $x - b$ is a factor of $P(x)$ and conversely, if $x - b$ is a factor of $P(x)$, then b is a zero of $y = P(x)$.

- The multiplicity of zero b is given by the highest power of $x - b$ that is a factor of $P(x)$.

- If b is a real number zero with multiplicity n of $y = P(x)$, then the graph of $y = P(x)$ crosses the x-axis at $x = b$ if n is odd, while the graph turns around and stays on the same side of the x-axis at $x = b$ if n is even.

- *Fundamental theorem of algebra:* Every polynomial function of degree $n \geq 1$ with complex number coefficients has at least one complex zero.

- *Number of zeros theorem:* Every polynomial function of degree $n \geq 1$ has exactly n complex zeros, where zeros of multiplicity k are counted k times.

- *Conjugate-pair theorems:*

  1. If a complex number $a + bi$ is a zero of a polynomial function of degree $n \geq 1$ with *real* number coefficients, then its conjugate $a - bi$ is also a zero.

4.4

A. Find the zeros of the function $P(x) = 2x^3(x + 4)^2(x - 3)^2$. Also, state the multiplicity of each zero and the degree of the polynomial.

B. Write a 5th degree polynomial function (in factored form) with zeros of 0, $-3 \pm i$, and $2 \pm \sqrt{5}$.

2. Let a, b, and c be rational numbers. If an irrational number $a + b\sqrt{c}$ is a zero of a polynomial function of degree $n \geq 1$ with *rational* number coefficients, then $a - b\sqrt{c}$ is also a zero.

Additional comments

- Be careful to note the type of number stated in a theorem about zeros of a polynomial function. You need a good grasp of number systems to really understand these statements, and it might be helpful to review Sections 1.1 and 2.2 concerning the relationships among the various types of numbers.

- The fundamental theorem of algebra is central to the theory of polynomial equations because it guarantees that every polynomial equation involving complex numbers can be solved using only complex numbers.

- When graphing a polynomial function, you may find it useful to remember that the number of turning points is at most one less than the degree of the polynomial function.

Detailed solutions to selected exercises

Exercise 4 If $P(x) = 4x^3 (x - 1)^5(x + 6)$, find the zeros of the function. Also, state the multiplicity of each zero and the degree of the polynomial function.

Solution From the factor theorem since x, $x - 1$, and $x + 6$ are factors, 0, 1, and -6 are zeros of the function. Since the factors are the third power of x, the fifth power of $x - 1$, and the first power of $x + 6$, we say

$$0 \text{ is a zero of multiplicity } 3$$
$$1 \text{ is a zero of multiplicity } 5$$
$$-6 \text{ is a zero of multiplicity } 1.$$

The degree of the polynomial function is given by the sum of the multiplicities of the zeros. Thus, the polynomial function is of degree 9.
Now do **A.**

Exercise 8 Write a 5th degree polynomial function (in factored form) with zeros of 5, $2 \pm i$, and $1 \pm \sqrt{3}$.

Solution By the factor theorem, if b is a zero of $P(x)$, then $x - b$ is a factor of $P(x)$. So, using the zeros given, we have

$$P(x) = (x - 5)[x - (2 + i)][x - (2 - i)][x - (1 + \sqrt{3})][x - (1 - \sqrt{3})].$$

Now do **B.**

Exercise 16 Graph the function $y = x^4 - 3x^2 + 2$ based on information obtained from the intercepts.

Solution By setting $x = 0$, we find that $(0,2)$ is the y-intercept. To find the x-intercepts, factor the polynomial as follows:

$$y = x^4 - 3x^2 + 2 = (x^2 - 2)(x^2 - 1)$$
$$= (x - \sqrt{2})(x + \sqrt{2})(x - 1)(x + 1).$$

Thus, the x-intercepts are $(\sqrt{2},0)$, $(-\sqrt{2},0)$, $(1,0)$, and $(-1,0)$. Since these intercepts are derived from factors with odd powers, the graph crosses the x-axis at each of these points. Using this information and also noting that large values of x produce large values of y, we draw the graph in Figure 4.4:16.

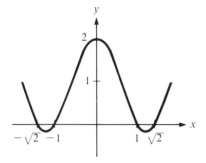

Figure 4.4:16

Now do **C**.

Exercise 26   Write a polynomial function (in factored form) with rational coefficients of the lowest possible degree with zeros of $3i$ and $1 + \sqrt{2}$.

Solution   Since the polynomial has rational coefficients, if $3i$ is a zero, then its conjugate $-3i$ is a zero, and if $1 + \sqrt{2}$ is a zero, then $1 - \sqrt{2}$ is also a zero. Thus, the lowest possible degree for the polynomial is 4 and one possibility for the function is

$$P(x) = (x - 3i)(x + 3i)[x - (1 + \sqrt{2})][x - (1 - \sqrt{2})].$$

The other possibilities for $P(x)$ include various constant factors that do not affect the zeros of the function.
Now do **D**.

Exercise 32   If $\sqrt{3}$ and $-4i$ are zeros of $P(x) = x^4 + 13x^2 - 48$, find the other zeros.

Solution   Since the polynomial function is degree 4, there are at most four distinct zeros. Because the coefficients of $P(x)$ are rational numbers, zeros of $\sqrt{3}$ and $-\sqrt{3}$ come in a pair, as do $-4i$ and $4i$. Thus, the four zeros are determined with the "other zeros" being $-\sqrt{3}$ and $4i$.
Now do **E**.

Exercise 36   If 1 and $-\frac{1}{2}$ are zeros of $P(x) = 2x^4 - 5x^3 + 11x^2 - 3x - 5$, then find the other zeros.

Solution   Since the polynomial function is degree 4, there are two remaining zeros. If 1 and $-\frac{1}{2}$ are zeros, we can use the factor theorem and synthetic division to factor $P(x)$ as follows:

$$
\begin{array}{r|rrrrr}
1\rfloor & 2 & -5 & 11 & -3 & -5 \\
 & & 2 & -3 & 8 & 5 \\
\hline
 & 2 & -3 & 8 & 5 & 0.
\end{array}
$$

Using the new quotient and the zero $-\frac{1}{2}$, we have

$$
\begin{array}{r|rrrr}
-\frac{1}{2}\rfloor & 2 & -3 & 8 & 5 \\
 & & -1 & 2 & -5 \\
\hline
 & 2 & -4 & 10 & 0.
\end{array}
$$

Thus, in factored form

$$P(x) = (x - 1)(x + \tfrac{1}{2})(2x^2 - 4x + 10).$$

C. Graph the function
$y = x^3 - 3x^2 - 4x$ based on
information obtained from the
intercepts.

D. Write a polynomial function (in
factored form) with rational
coefficients of the lowest possible
degree with zeros of $1 + 2i$ and
$\sqrt{3}$.

E. If $2 - i$ and $\sqrt{2}$ are zeros of $P(x)$
$= x^4 - 4x^3 + 3x^2 + 8x - 10$,
find the other zeros.

67

68

F. If 2 and $-\frac{1}{2}$ are zeros of $P(x)$ $= 4x^4 - 8x^3 + x^2 - x - 2$, then find the other zeros.

Setting the factor $2x^2 - 4x + 10 = 0$, we simplify to $x^2 - 2x + 5 = 0$ and find the other two zeros by using the quadratic formula.

$$x = \frac{-(-2) \pm \sqrt{(-2)^2 - 4(1)(5)}}{2(1)} = \frac{2 \pm \sqrt{-16}}{2} = \frac{2 \pm 4i}{2} = 1 \pm 2i$$

Thus, $1 + 2i$ and $1 - 2i$ are the remaining zeros.
Now do **F.**

4.5 Rational zeros and the location theorem

Objectives checklist

Can you:

a. List the possible rational zeros of a polynomial function with integer coefficients?

b. Use Descartes' rule of signs to determine the maximum number of positive and negative real zeros of a polynomial function?

c. Find all the zeros in a polynomial function when given a polynomial with a sufficient number of rational zeros so all the zeros can be determined?

d. Use the location theorem to verify the existence of a zero of a function in a given interval?

e. Approximate the zero of a function in a given interval to the nearest tenth?

Key terms

Upper bound   Lower bound

Key rules and formulas

- *Rational zero theorem:* If p/q, a rational number in lowest terms, is a zero of the polynomial function with integer coefficients

$$P(x) = a_nx^n + a_{n-1}x^{n-1} + \cdots + a_1x + a_0 \qquad (a_n \neq 0),$$

  p is an integral factor of the constant term a_0 and q is an integral factor of the leading coefficient a_n.

- *Descartes' rule of signs:* The maximum number of positive real zeros of the polynomial function $y = P(x)$ is the number of changes in sign of the coefficients in $P(x)$. The number of changes in sign of the coefficients in $P(-x)$ is the maximum number of negative real zeros. In both cases, if the number of zeros is not the maximum number, then it is less than this number by a multiple of 2.

- *Upper and lower bounds theorem*

  1. If we divide a polynomial $P(x)$ synthetically by $x - b$, where $b > 0$, and all the numbers in the bottom row have the same sign, then there is no zero greater than b. We say b is an upper bound for the zeros of $y = P(x)$.
  2. If we divide a polynomial $P(x)$ synthetically by $x - c$, where $c < 0$, and the numbers in the bottom row alternate in sign, then there is no zero less than c. We say c is a lower bound for the zeros of $y = P(x)$.

  For the purposes of these tests, zero may be denoted as $+0$ or -0.

- *Location theorem:* Let $y = P(x)$ be a polynomial function. If $P(a)$ and $P(b)$ have opposite signs, then $y = P(x)$ has at least one real zero between a and b.

- To approximate an irrational zero known to exist in a given interval, repeatedly halve the interval and use the location theorem until a specified degree of accuracy is attained.

Detailed solutions to selected exercises

Exercise 4   List the possible rational zeros of $P(x) = 3x^3 + 7x^2 + 8$.

Solution   By the rational zero theorem, if p/q is a zero of $y = P(x)$, then p must be an integer that is a factor of 8, while q must be an integral factor of 3. Thus, the possibilities are

$$p = \pm 8, \pm 4, \pm 2, \pm 1; \quad q = \pm 3, \pm 1$$

and the possible rational zeros p/q are

$$\pm 8, \pm 4, \pm \tfrac{8}{3}, \pm 2, \pm \tfrac{4}{3}, \pm 1, \pm \tfrac{2}{3}, \pm \tfrac{1}{3}.$$

Now do **A**.

Exercise 8   Use Descartes' rule of signs to determine the maximum number of positive and negative real zeros for $P(x) = x^8 + 1$.

Solution   Both coefficients in $P(x)$ are positive. Since there are no sign changes in the coefficients, there are no positive real zeros for $y = P(x)$. Then for $P(-x)$ we have

$$P(-x) = (-x)^8 + 1 = x^8 + 1.$$

The coefficients in $P(-x)$ also do not change sign, so there are no negative real zeros. In this case the eight zeros are complex numbers.
Now do **B**.

Exercise 12   Find the zeros of $P(x) = 4x^3 - x^2 - 28x + 7$.

Solution

1. Use Descartes' rule of signs to determine the maximum number of positive and negative real zeros.

$$P(x) = 4x^3 - x^2 - 28x + 7$$
$$P(-x) = -4x^3 - x^2 + 28x + 7$$

   There are two sign changes in the coefficients of $P(x)$ and one sign change in the coefficients of $P(-x)$. Therefore, at most there may be two positive real zeros and there is exactly one negative real zero.

2. Determine the possible rational zeros. By the rational zero theorem, the choices are

$$\frac{p}{q} = \pm \frac{7,1}{4,2,1} = 7, \tfrac{7}{2}, \tfrac{7}{4}, 1, \tfrac{1}{2}, \tfrac{1}{4}, -\tfrac{1}{4}, -\tfrac{1}{2}, -1, -\tfrac{7}{4}, -\tfrac{7}{2}, -7.$$

3. Use synthetic division to test the possibilities. We'll start with the positive choices since there may be two positive answers. Let's pick 1. This number is easy to evaluate and since it's in the middle of the positive choices, if it's not a zero, maybe it's an upper bound that eliminates other possibilities.

$$
\begin{array}{r|rrrr}
1\rfloor & 4 & -1 & -28 & 7 \\
 & & 4 & 3 & -25 \\
\hline
 & 4 & 3 & -25 & -18
\end{array}
$$

4.5

A. List the possible rational zeros of $P(x) = 6x^4 + 8x^3 + 5$.

B. Use Descartes' rule of signs to determine the maximum number of positive and negative real zeros for $P(x) = x^7 + 2$.

C. Find the zeros of
$P(x) = 2x^3 + x^2 - 10x - 5.$

No luck this time, so we keep going. By trial and error, we'll eventually come to $\frac{1}{4}$.

$$
\begin{array}{r|rrrr}
\frac{1}{4} & 4 & -1 & -28 & 7 \\
& & 1 & 0 & -7 \\
\hline
& 4 & 0 & -28 & 0
\end{array}
$$

Thus, $\frac{1}{4}$ is a zero and $P(x)$ can be factored as

$$P(x) = (x - \tfrac{1}{4})(4x^2 - 28).$$

The remaining zeros of $y = P(x)$ are the solutions of $4x^2 - 28 = 0$.

$$4x^2 - 28 = 0 \text{ so } x^2 = 7 \text{ so } x = \pm\sqrt{7}$$

D. Given that the function
$P(x) = x^3 + 3x^2 - 3x - 7$ has
a zero between 1 and 2,
approximate the zero to the
nearest tenth.

Thus, the three zeros are $\frac{1}{4}$, $\sqrt{7}$, and $-\sqrt{7}$.
Now do **C.**

Exercise 20   Use the location theorem to verify that
$P(x) = 4x^3 - x^2 - 28x + 7$ has a zero between 2 and 3.

Solution   By direct substitution or synthetic division, we find that
$P(2) = -21$ and $P(3) = 22$. Since $P(2)$ and $P(3)$ have opposite signs, the
location theorem tells us that at least one real zero lies between 2 and 3.

Exercise 26   Given that $P(x) = x^3 - 3x + 1$ has a zero between 1 and 2,
approximate the zero to the nearest tenth.

Solution   Using the location theorem and repeatedly halving the interval
containing the zero, we determine that

$$
\begin{aligned}
P(1) &= -1 \text{ and } P(2) = 3, \text{ so a zero is between 1 and 2} \\
P(1.5) &= -0.125, \text{ so a zero is between 1.5 and 2} \\
P(1.75) &= 1.11, \text{ so a zero is between 1.5 and 1.75} \\
P(1.63) &= 0.441, \text{ so a zero is between 1.5 and 1.63} \\
P(1.57) &= 0.160, \text{ so a zero is between 1.5 and 1.57} \\
P(1.54) &= 0.032, \text{ so a zero is between 1.5 and 1.54.}
\end{aligned}
$$

Thus to the nearest tenth, the zero is 1.5.
Now do **D.**

4.6 Rational functions

Objectives checklist

Can you:

a. Find any vertical asymptotes for a rational function?

b. Determine the behavior of the graph of a rational function near any
vertical asymptote?

c. Find any horizontal asymptotes for a rational function and determine any
points at which the curve crosses the horizontal asymptote?

d. Find any oblique asymptotes for a rational function?

e. Find any point at which there is a hole in the graph of a rational function?

f. Graph a rational function?

Key terms

Rational function   Horizontal asymptote
Vertical asymptote   Oblique asymptote

Key rules and formulas

- *Vertical asymptotes:* If $P(x)$ and $Q(x)$ have no common factors, the rational function $y = P(x)/Q(x)$ has as vertical asymptote the line $x = a$ for each value a at which $Q(a) = 0$.

- If $P(x)$ and $Q(x)$ have no common factors, and if $(x - a)^n$ is a factor of $Q(x)$, where n is the largest positive integer for which this statement is true, then:

  1. The graph of $y = P(x)/Q(x)$ goes in opposite directions about the vertical asymptote $x = a$ when n is odd.
  2. The graph of $y = P(x)/Q(x)$ goes in the same direction about the vertical asymptote $x = a$ when n is even.

- *Horizontal asymptote theorem:* The graph of the rational function

$$y = \frac{P(x)}{Q(x)} = \frac{a_n x^n + a_{n-1} x^{n-1} + \cdots + a_0}{b_m x^m + b_{m-1} x^{m-1} + \cdots + b_0},$$

where $a_n, b_m \neq 0$, has

  1. a horizontal asymptote at $y = 0$ (the x-axis) if $n < m$
  2. a horizontal asymptote at $y = a_n/b_m$ if $n = m$
  3. no horizontal asymptote if $n > m$

- *Oblique asymptotes:* The graph of $y = P(x)/Q(x)$ has an oblique asymptote if the degree of $P(x)$ is one more than the degree of $Q(x)$. We determine an oblique asymptote by dividing $P(x)$ by $Q(x)$ and setting y equal to the quotient.

Additional comments

- An outline of the procedure for graphing rational functions with the form $y = P(x)/Q(x)$ appears in the text at the end of Section 4.6.

Detailed solutions to selected exercises

Exercise 2 Find any vertical asymptotes for $y = \dfrac{x^2 + 1}{x(x + 4)}$.

Solution $P(x)$ and $Q(x)$ have no common factors and the polynomial in the denominator $Q(x) = x(x + 4)$ equals 0 when x is 0 or -4. Thus, there are two vertical asymptotes: $x = 0$ and $x = -4$.
Now do **A**.

Exercise 10 Determine any asymptotes for $y = (x + 3)/(x - 5)$ and graph the function.

Solution We follow the steps outlined in the text.

  1. $P(x)$ and $Q(x)$ have no common factors so continue to Step 2.
  2. The polynomial in the denominator, $x - 5$, equals zero when x is 5. Thus, $x = 5$ is a vertical asymptote. Since the degree of $x - 5$ is 1, which is odd, the graph goes in opposite directions about the vertical asymptote.
  3. $P(x)$ and $Q(x)$ are both first-degree polynomials. By the horizontal asymptote theorem, if $n = m$, the horizontal asymptote is $y = a_n/b_m$. In this case both these coefficients are 1, so the horizontal asymptote is $y = 1$.

4.6

A. Find all vertical asymptotes for
$$y = \frac{x^2(x - 6)}{(x + 3)(x - 8)}.$$

B. Determine any asymptotes for $y = \dfrac{x + 1}{x}$ and graph the function.

4. To determine if the curve ever crosses the horizontal asymptote, we set y equal to 1 and solve

$$1 = \frac{x + 3}{x - 5}.$$

This equation is equivalent to $x - 5 = x + 3$, which is never true. Therefore, the curve does not cross the horizontal asymptote.

5. When $x = 0$, $y = (0 + 3)/(0 - 5) = -\frac{3}{5}$ so the y-intercept is $(0, -\frac{3}{5})$. By setting $y = 0$, we solve $0 = x + 3$ to determine that the x-intercept is $(-3, 0)$.

6. Using the above information and plotting one point to the right of the vertical asymptote, say $(7, 5)$, we draw the graph in Figure 4.6:10.

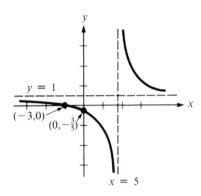

Figure 4.6:10

Now do **B.**

Exercise 24   Determine any asymptotes for $y = \dfrac{x^2 - 2x - 3}{x + 3}$ and graph the function.

Solution   We follow the steps outlined in the text.

1. $P(x)$ and $Q(x)$ have no common factors so continue to Step 2.
2. $Q(x)$ equals zero when $x = -3$. Thus, the line $x = -3$ is a vertical asymptote. Also, since the degree of $x + 3$ is 1, which is odd, the graph goes in opposite directions about $x = -3$.
3. The degree of the numerator is one more than the degree of the denominator, so there is no horizontal asymptote but there is an oblique asymptote. Since

$$y = \frac{x^2 - 2x - 3}{x + 3} = x - 5 + \frac{12}{x + 3},$$

the oblique asymptote is the line $y = x - 5$.
4. There is no horizontal asymptote so continue to Step 5.

5. By setting $x = 0$, we may determine that $(0, -1)$ is the y-intercept. Similarly, by setting $y = 0$, we may determine that $(3,0)$ and $(-1,0)$ are x-intercepts.

6. Using the above information and plotting a point to the left of the vertical asymptote, say $(-5, -16)$, we graph the function as shown in Figure 4.6:24.

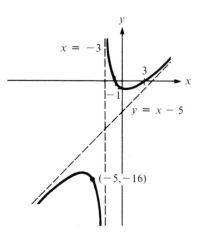

Figure 4.6:24

C. Determine any asymptotes for
$$y = \frac{x^2 + 1}{x}$$
and graph the function.

Now do **C.**

Exercise 32   Graph $y = x/(x^2 - 2x)$.

Solution   We follow the steps outlined in the text.

1. $P(x)$ and $Q(x)$ have a common factor of x, which we divide out as follows:

$$y = \frac{x}{x^2 - 2x} = \frac{x}{x(x - 2)} = \frac{1}{x - 2} \qquad \text{if } x \neq 0.$$

Since the common factor is of the same degree in the numerator and the denominator, there is a hole in the graph at $x = 0$.

2. At this point, we only consider the function $y = 1/(x - 2)$. The denominator in this function is 0 when x is 2, so the vertical asymptote is $x = 2$. Since the degree of $x - 2$ is 1, which is odd, the graph goes in opposite directions around the vertical asymptote.

3. The polynomial in the denominator is of larger degree than the polynomial in the numerator, so the horizontal asymptote is $y = 0$ (the x-axis).

4. The curve never crosses the horizontal asymptote since there are no values of x for which $0 = 1/(x - 2)$.

5. Since y cannot be 0, there is no x-intercept. Although the y-intercept for $y = 1/(x - 2)$ is $(0, -\frac{1}{2})$, there is no y-intercept for $y = x/(x^2 - 2x)$ since the function is undefined at $x = 0$.

74

D. Graph $y = \dfrac{x^2 - 4}{x + 2}$.

6. Using the above information and plotting points to the left and right of the vertical asymptote, say $(1, -1)$ and $(3, 1)$, we draw the graph in Figure 4.6:32.

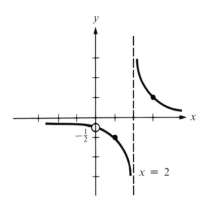

Figure 4.6:32

Now do **D.**

Sample test questions: Chapter 4

1. Complete the statement.

   a. The range of $y = 1 - 6x - x^2$ is _____ .

   b. The degree of the polynomial function $y = x(x - 3)^2(x - 2)^3$ is _____ .

   c. The remainder when $x^3 + 27$ is divided by $x - 3$ is _____ .

   d. If $2 + \sqrt{7}$ is a zero of $y = x^2 - 4x - 3$, the other zero is _____ .

   e. The vertical asymptote of $y = x/(1 - x)$ is _____ .

   f. The horizontal asymptote of $y = x/(1 - x)$ is _____ .

   g. By Descartes' rule of signs, the maximum number of positive real zeros for $P(x) = 2x^4 + 7x^3 + 25x^2 + 47x + 18$ is _____ .

   h. A polynomial function (in factored form) with zeros of 2, -3, and 0 is

   _____ .

   i. The x-intercepts of $y = x^2 - x - 8$ are _____ .

   j. By the rational zero theorem, the possible rational zeros of $P(x) = 3x^3 + 16x^2 - 2$ are _____ .

   k. The slope of the line $(x/4) + (y/5) = 1$ is _____ .

   l. The slope of the line perpendicular to $y - 3x = 5$ is _____ .

2. Write the equation of the linear function f if $f(-1) = 4$ and $f(5) = -3$.

3. Find the zeros of the function or the roots of the equation.

   a. $y = 2x^2 - 3$ b. $y = (x^2 - 1)/x^3$

   c. $x^3 - 4x^2 + x + 6 = 0$

4. Solve each inequality.

    a. $x^2 - x \le 12$ b. $2x^2 + 7 \ge 0$ c. $2 < x^2$

5. a. Divide $x^4 - 6x + 1$ by $x^2 - 2$ using long division.

    b. Divide $3x^4 - 2x^3 + x - 7$ by $x + 1$ using synthetic division.

6. If $P(x) = 3x^4 - x^3 + 7$, find $P(-3)$ by (a) direct substitution and (b) the remainder theorem.

7. If -1 is a zero of $y = 6x^3 + 5x^2 - 3x - 2$, then find the other zeros.

8. Given that $P(x) = 2x^3 - 3x^2 - 12x + 6$ has a zero between 3 and 4, approximate the zero to the nearest tenth.

9. a. Determine the vertical asymptote(s) for $y = \dfrac{x - 1}{x(x - 1)(x + 7)}$.

    b. Determine the horizontal asymptote for $y = (5x - 3)/(x^2 - 3x)$. Find the points (if any) where the curve crosses the horizontal asymptote.

10. Solve each applied problem.

    a. What is the maximum product for two numbers whose average is 20? (An algebraic solution is required.)

    b. The height (y) of a projectile shot vertically up from the ground with an initial velocity of 96 ft/sec is given by the formula $y = 96t - 16t^2$. For what values of t is the projectile more than 80 ft off the ground?

11. Graph the given function.

    a. $y = x^2 - 4x + 7$ b. $y = 2/(x + 1)$ c. $h(x) = (x^2 + x)/x$

12. Select the choice that completes the statement or answers the question.

    a. Is the function $y = 1/x$ a polynomial function?

        (a) Yes   (b) No

    b. If $P(x) = x(x - 1)^3(x + 1)^2$, what is the multiplicity of the zero of -1?

        (a) 1   (b) 2   (c) 3   (d) 5   (e) 6

    c. Which number is a zero for $P(x) = 2x^3 + 5x^2 + 4x + 1$?

        (a) $\frac{1}{2}$ (b) -2 (c) 4   (d) -1 (e) 1

    d. The lowest degree for a polynomial function with *real* coefficients with known zeros of $1 - i$ and $\sqrt{3}$ is

        (a) 2   (b) 3   (c) 4   (d) 5   (e) 6

    e. The equation of the line through $(2, -3)$ whose slope is undefined is

        (a) $x = 2$ (b) $y = -3$ (c) $y = 2x - 3$

        (d) $y = x - 3$ (e) $x = -3$

Chapter 5
Exponential and Logarithmic Functions

5.1 Exponential functions

Objectives checklist

Can you:

a. Determine the missing component in an ordered pair so that the pair is an element in a given exponential function?

b. Graph an exponential function and determine the range of the function?

c. Find the base of the exponential function $y = b^x$ that contains a given point?

d. Solve applied problems involving exponential functions whose domain is the set of nonnegative integers?

Key term

Exponential function

Key rules and formulas

- The function f defined by $f(x) = b^x$, where $b > 0$ with $b \neq 1$, is called the exponential function with base b.

  Domain of f: $(-\infty, \infty)$
  Range of f: $(0, \infty)$

- The x-axis is a horizontal asymptote for the graph of $y = b^x$ ($b > 0$, $b \neq 1$). There are no x-intercepts and the y-intercept is $(0,1)$.

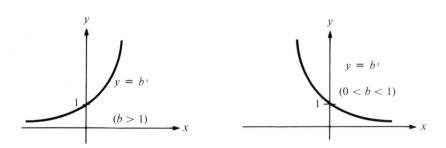

- For $y = b^x$, if $b > 1$, then as x increases, y increases and the function is useful in applications where y is growing exponentially (see Figure).

- For $y = b^x$, if $0 < b < 1$, then as x increases, y decreases and the function is useful in applications where y is decaying or depreciating exponentially (see Figure).

- For $b > 0$, $b \neq 1$, and any real number x, if x lies between the rational numbers r and s, then b^x lies between b^r and b^s.

- All previous laws of exponents are valid for all real number exponents.

- Since exponential functions are one-to-one, if b is a positive number other than 1, then

$$b^x = b^y \text{ implies } x = y.$$

Additional comments

- Numbers with irrational exponents, such as 2^π, are real numbers. Approximate values for these numbers may be found by using the $\boxed{y^x}$ key on a calculator.

- A scientific calculator is particularly useful throughout this chapter.

Detailed solutions to selected exercises

Exercise 4   If $g(x) = 4^{-x}$, fill in the missing component in each ordered pair: $(2,\ )$, $(\ ,0.5)$, $(-0.5,\ )$, $(\ ,4)$. Also, graph the function.

Solution   Since $4^{-x} = (\frac{1}{4})^x$, we may work with $g(x) = 4^{-x}$ or $g(x) = (\frac{1}{4})^x$. It is easiest to find y when given x. So,

if $x = 2$, $y = 4^{-2} = \frac{1}{16}$, thus the pair $(2,\frac{1}{16})$

if $x = -0.5$, $y = 4^{-(-0.5)} = 4^{1/2} = \sqrt{4} = 2$, thus the pair $(-0.5,2)$.

We find x when given y either by inspection or by applying the principle that $b^x = b^y$ implies $x = y$.

if $y = 0.5 = 4^{-x}$, then $2^{-1} = 2^{-2x}$ so $x = \frac{1}{2}$, thus the pair $(\frac{1}{2},0.5)$

if $y = 4 = 4^{-x}$, then $x = -1$ (by inspection), thus the pair $(-1,4)$

The function is graphed in Figure 5.1:4.

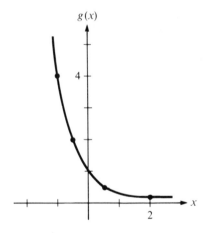

Figure 5.1:4

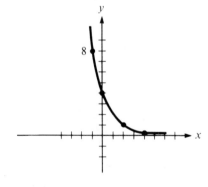

Figure 5.1:16

Now do **A**.

Exercise 16   Graph $y = 2^{2-x}$.

Solution   By plotting a few convenient points, say $(0,4)$, $(2,1)$ $(4,\frac{1}{4})$, and $(-1,8)$ and knowing that the x-axis is a horizontal asymptote for the graph, we can draw the graph shown in Figure 5.1:16.

We can also obtain this graph through analysis by noting that 2^{2-x} is equivalent to $2^{-(x-2)}$, so that the graph of $y = 2^{2-x}$ may be obtained by shifting the graph of $y = 2^{-x}$ [see text Figure 5.1(b)] 2 units to the right. We read from the graph that the range of the function $y = 2^{2-x}$ is $(0,\infty)$.

Now do **B**.

Chapter 5

5.1

A. If $g(x) = 3^{-x}$, fill in the missing component in each ordered pair: $(2,\ )$, $(\ ,\frac{1}{3})$, $(-\frac{1}{3},\ )$, $(\ ,3)$.

B. Graph $y = 2^{x-2}$.

C. If $2^x = 0.315$, find an integer n so that $n < x < n + 1$.

D. A biologist grows a colony of bacteria. It is found experimentally that $N = N_0 2^t$, where N represents the number of bacteria present at the end of t days. N_0 is the number of bacteria present at the start of the experiment. Suppose that there are 104,000 bacteria present at the end of 3 days.
  a. How many bacteria were present at the start of the experiment?
  b. How many bacteria are present at the end of 5 days?
  c. At the end of how many days are there 832,000 bacteria present?

E. You purchase a home for $180,000 and take out a 20-year mortgage. If the annual inflation rate remains fixed at 6% and the value of the home keeps pace with inflation, then what is the value of your home by the time the mortgage is paid off?

Exercise 22c If $10^x = 0.451$, find an integer n so $n < x < n + 1$.

Solution $10^0 = 1$ and $10^{-1} = 0.1$. Since 0.451 is between 0.1 and 1, 10^x is between 10^{-1} and 10^0, so $n = -1$.
Now do **C**.

Exercise 28 See text for question.

Solution

a. Substituting 2 for t and 153,000 for N in the given formula, we have
$$153,000 = N_0 3^2 \quad \text{so} \quad N_0 = \tfrac{153,000}{9} = 17,000.$$

b. Since $N_0 = 17,000$, the formula is $N = 17,000(3)^t$. If $t = 4$,
$$N = 17,000(3)^4 = 1,377,000.$$

c. We find t when $N = 459,000$ by solving
$$459,000 = 17,000(3)^t \quad \text{so} \quad 27 = 3^t \quad \text{so} \quad t = 3.$$
Thus, $t = 3$ days.

Now do **D**.

Exercise 30 You purchase a home for $150,000 and take out a 30-year mortgage. If the annual inflation rate remains fixed at 5 percent and the value of the home keeps pace with inflation, then what is the value of your home by the time the mortgage is paid off?

Solution We can consider the $150,000 to be an investment compounded annually at the rate of 5 percent for 30 years. By considering Example 7 in the text, we can determine that the general formula for an investment that is compounded annually is $A = P(1 + r)^t$. Substituting $P = 150,000$, $r = 0.05$, and $t = 30$ into this formula gives $A = 648,291.36$. Thus, the home is worth $648,291.36.
Now do **E**.

5.2 Logarithmic functions

Objectives checklist

Can you:

a. Convert an exponential statement in the form $b^y = x$ to its equivalent logarithmic statement $\log_b x = y$, and vice versa?

b. Determine the value of an unknown in certain types of simple logarithmic statements?

c. Graph a logarithmic function?

d. Find the domain of a logarithmic function?

e. Determine the common logarithm and antilogarithm of a number by using a calculator?

f. Solve applied problems involving logarithmic functions?

Key terms

Logarithm
Common logarithm
Logarithmic function

Natural logarithm
Antilogarithm (inverse logarithm)

Key rules and formulas

- The logarithm (or log) of a number is the exponent to which a fixed base is raised to obtain the number. The statements

$$\log_b N = L \text{ and } b^L = N$$

are equivalent, and we say L is the log to the base b of N.

- For $b > 0$, $b \neq 1$, and $x > 0$,

$$y = \log_b x \text{ if and only if } x = b^y.$$

- The exponential function $y = b^x$ and the logarithmic function $y = \log_b x$ are inverse functions and are graphed below. Note the following:

  1. The domain of $y = \log_b x$ (which is the range of $y = b^x$) is $(0,\infty)$. This means we may only take the log of a positive number.
  2. The range of $y = \log_b x$ (which is the domain of $y = b^x$) is $(-\infty,\infty)$.
  3. The graph of $y = \log_b x$ has the x-intercept $(1,0)$ and the y-axis is a vertical asymptote.
  4. The graphs of $y = \log_b x$ and $y = b^x$ are reflections of each other across the line $y = x$.

A. Express $(\frac{1}{3})^{-4} = 81$ in logarithmic form.

B. Express $\log_{1/5} 25 = -2$ in exponential form.

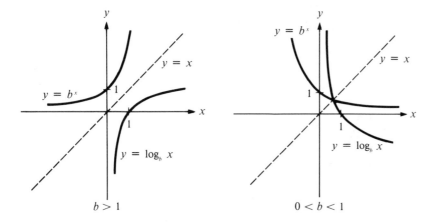

$b > 1$ ⠀⠀⠀⠀⠀⠀⠀ $0 < b < 1$

Additional comments

- To determine base 10 or common logarithms by calculator, simply enter the number and press the $\boxed{\log}$ key. To determine an antilogarithm, enter the number, then press the sequence $\boxed{\text{INV}}$ $\boxed{\log}$.

Detailed solutions to selected exercises

Exercise 6 Express $(\frac{1}{2})^{-3} = 8$ in logarithmic form.

Solution $(\frac{1}{2})^{-3} = 8$ is equivalent to $\log_{1/2} 8 = -3$ and we say -3 is the log to the base $\frac{1}{2}$ of 8.
Now do **A.**

Exercise 16 Express $\log_{1/4} 4 = -1$ in exponential form.

Solution The statement $\log_{1/4} 4 = -1$ tells us that placing the exponent -1 above a base of $\frac{1}{4}$ results in the number 4, so $\log_{1/4} 4 = -1$ is equivalent to $(\frac{1}{4})^{-1} = 4$.
Now do **B.**

C. Evaluate $\log_8 2$.

D. If $\log_b 10 = \frac{1}{3}$, find b.

E. If $g(x) = \log_4 x$, find $g(16)$, $g(\frac{1}{4})$, and $g(32)$.

F. Graph $y = \log_{1/4} x$.

Exercise 28   Evaluate $\log_9 3$.

Solution $\log_9 3$ is the number c satisfying $9^c = 3$. Since $9^{1/2} = \sqrt{9} = 3$, $c = \log_9 3 = \frac{1}{2}$.
Now do **C**.

Exercise 36   If $\log_b 10 = \frac{1}{2}$, find b.

Solution $\log_b 10 = \frac{1}{2}$ is equivalent to $b^{1/2} = 10$. Squaring both sides of this equation gives $b = 100$.
Now do **D**.

Exercise 42   If $g(x) = \log_9 x$, find $g(3)$, $g(27)$, and $g(\frac{1}{81})$.

Solution   Replacing x by the given values, we have $g(3) = \log_9 3 = \frac{1}{2}$ (see solution to Exercise 28). Next, $g(27) = \log_9 27 = c$. Then $9^c = 27$, so $3^{2c} = 3^3$ and $c = \frac{3}{2}$. Finally, $g(\frac{1}{81}) = \log_9 \frac{1}{81} = c$. Then $9^c = \frac{1}{81}$, so $9^c = 9^{-2}$ and $c = -2$.
Now do **E**.

Exercise 46   Graph $y = \log_{1/2} x$.

Solution   The basic graph for $y = \log_b x$ when $0 < b < 1$ was shown above. Specifically, if $b = \frac{1}{2}$, set up a table of values as below (remember x must be positive) and graph the function as in Figure 5.2:46.

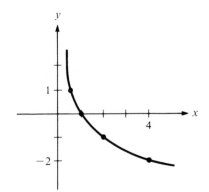

Figure 5.2:46

x	y
1	$\log_{1/2} 1 = 0$
2	$\log_{1/2} 2 = -1$
4	$\log_{1/2} 4 = -2$
$\frac{1}{2}$	$\log_{1/2} \frac{1}{2} = 1$

Now do **F**.

Exercise 52   Graph $y = 1 + \log_2 x$.

Solution   We start with the graph of $y = \log_2 x$ which can be obtained from a table of values (see Figure 5.3 in the text). To graph $y = 1 + \log_2 x$, we shift the graph of $y = \log_2 x$ up 1 unit. The completed graph is shown in Figure 5.2:52.

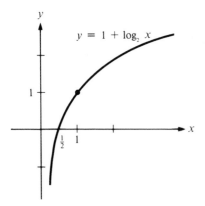

Figure 5.2:52

G. Graph $y = 1 - \log_2 x$.

H. What is the domain of
$f(x) = \log_5 \sqrt{x - 2}$?

Now do **G.**

Exercise 60 What is the domain of $f(x) = \log_{10} \sqrt{x + 1}$?

Solution Since logarithms are defined only for positive numbers, $\sqrt{x + 1}$ must be greater than 0. Thus, $x + 1 > 0$, so $x > -1$, and the domain is $(-1, \infty)$.

I. Determine the pH for a solution with $[H^+] = 1.6 \times 10^{-5}$.

Now do **H.**

Exercise 64 Determine the pH for swimming pool water given $[H^+] = 3.3 \times 10^{-8}$.

Solution By the formula given in the text,

$$pH = -\log[H^+] = -\log(3.3 \times 10^{-8}) = 7.5.$$

By calculator, the keystroke sequence is 3.3 $\boxed{\text{EE}}$ 8 $\boxed{+/-}$ $\boxed{\text{log}}$ $\boxed{+/-}$.
Now do **I.**

J. Find $[H^+]$ for a solution if its pH is 7.4.

Exercise 68 Find $[H^+]$ for household ammonia if its pH is 11.3.

Solution In the formula $pH = -\log [H^+]$, we replace pH by 11.3 and solve for $\log[H^+]$. Thus, $\log [H^+] = -11.3$. We now compute an inverse logarithm by the following calculator sequence:

11.3 $\boxed{+/-}$ $\boxed{\text{INV}}$ $\boxed{\text{log}}$ $\boxed{5.0119 - 12}$.

Thus, $[H^+] = 5.0 \times 10^{-12}$.
Now do **J.**

5.3 Properties of logarithms

Objectives checklist

Can you:

a. Use the properties of logarithms to express certain log statements as the sum or difference of simpler logarithms?

b. Use the properties of logarithms to express sums and differences of logarithms as a single logarithm with coefficient 1?

c. Use the properties of logarithms to simplify certain logarithmic expressions?

82

5.3

A. Express $\log_b(x/\sqrt{y})$ as the sum or difference of simpler logarithms.

B. Express $2\log_b x - \frac{1}{3}\log_b y$ as a single logarithm with coefficient 1.

C. Simplify $\log_3(81^{1/3})$.

D. Simplify $b^{\log_b x - \log_b y}$.

E. If $\log_a 2 = x$ and $\log_a 3 = y$, express $\log_a 48$ in terms of x and y.

Key rules and formulas

- *Properties of Logarithms:* If b, x_1, x_2 are positive, with $b \neq 1$ and k any real number, then

  1. $\log_b x_1 x_2 = \log_b x_1 + \log_b x_2$
  2. $\log_b(x_1/x_2) = \log_b x_1 - \log_b x_2$
  3. $\log_b(x_1)^k = k \log_b x_1$
  4. $\log_b b = 1$
  5. $\log_b 1 = 0$
  6. $\log_b b^x = x$
  7. $b^{\log_b x} = x$ (if $x > 0$)

Additional comments

- Since a logarithm is an exponent, the properties of logarithms may be derived from the properties of exponents.

Detailed solutions to selected exercises

Exercise 14 Express $\log_b\sqrt{x/y}$ as the sum or difference of simpler logarithms.

Solution
$\log_b\sqrt{x/y} = \log_b(x/y)^{1/2}$
$= \frac{1}{2}\log_b(x/y)$ Using property 3
$= \frac{1}{2}(\log_b x - \log_b y)$ Using property 2

Now do **A**.

Exercise 26 Express $3\log_b x - \frac{1}{2}\log_b z$ as a single log with coefficient 1.

Solution $3\log_b x - \frac{1}{2}\log_b z = \log_b x^3 - \log_b z^{1/2}$
$= \log_b(x^3/z^{1/2})$ or $\log_b(x^3/\sqrt{z})$

Now do **B**.

Exercise 34 Simplify $\log_2(4^{1/3})$.

Solution $\log_2(4^{1/3}) = \log_2(2^2)^{1/3} = \log_2 2^{2/3}$. Since $\log_b b^x = x$, we have $\log_2 2^{2/3} = \frac{2}{3}$.

Now do **C**.

Exercise 40 Simplify $b^{\log_b x + \log_b y}$.

Solution $b^{\log_b x + \log_b y} = b^{\log_b xy}$ Using property 1
$= xy$

Now do **D**.

Exercise 42b If $\log_a 2 = x$ and $\log_a 3 = y$, express $\log_a 72$ in terms of x and y.

Solution Using factors of 2 and 3, we write 72 as $2^3 \cdot 3^2$. Then

$\log_a 72 = \log_a(2^3 \cdot 3^2) = \log_a 2^3 + \log_a 3^2 = 3\log_a 2 + 2\log_a 3$.

Since $\log_a 2 = x$ and $\log_a 3 = y$, we have $\log_a 72 = 3x + 2y$.
Now do **E**.

Exercise 44d By counterexample, disprove $\dfrac{\log_b x}{\log_b y} = \log_b x - \log_b y$.

Solution If we replace x, y, and b by 10, then $(\log_{10} 10)/(\log_{10} 10) = 1/1 = 1$, while $\log_{10} 10 - \log_{10} 10 = 1 - 1 = 0$. Be careful not to confuse this incorrect statement with the second property of logarithms stated above.

Now do **F**.

Exercise 48 See text for question.

Solution $\log_{10}(\frac{1}{2}) = -0.3010$, so $\log_{10}(\frac{1}{2})$ is a negative number. Thus, when we multiply both sides of the inequality by $\log_{10}(\frac{1}{2})$, we must reverse the sense of the inequality.

5.4 Exponential and logarithmic equations

Objectives checklist

Can you:

a. Use logarithms to solve exponential equations and equations containing a difficult exponent?

b. Solve a logarithmic equation?

c. Change a logarithm in one base to a logarithm in another base?

Key terms

Exponential equation   Logarithmic equation

Key rules and formulas

*   The logarithm rule defines a function that is one-to-one. Thus, if x, y, and b are positive numbers (with $b \neq 1$), then

$$x = y \text{ implies } \log_b x = \log_b y$$

    and, conversely,

$$\log_b x = \log_b y \text{ implies } x = y.$$

*   Using the above rule, we solve exponential equations (the unknown is in the exponent) and equations containing a difficult exponent as follows:

    1. Take the logarithm of both sides of the equation.
    2. Simplify by using the property $\log_b x^k = k \log_b x$.
    3. Solve the resulting equation in the usual way.

*   We solve a logarithmic equation (the unknown is in the log statement) in some cases by using $\log_b x = \log_b y$ implies $x = y$, and in other cases by changing the equation from logarithmic form to exponential form.

*   *Change of base:* We change a logarithm in one base to a logarithm in another base as follows:

$$\log_b x = \frac{\log_a x}{\log_a b}.$$

Additional comments

*   Always check your solutions to logarithmic equations in the original equation and accept only solutions that result in the logarithms of positive numbers.

Detailed solutions to selected exercises

Exercise 6 Solve $3^{2x} = 5$.

F. By counterexample, disprove
$\log_b (x + y) = \log_b x \cdot \log_b y$.

5.4

A. Solve $5^{2x} = 7$.

B. Solve $x^{-2.4} = 58.3$.

C. Solve $\log(2x) = \frac{1}{3}$.

D. Solve $\log_5 x - \log_5 2 = 3$.

E. Solve
$\log_5(x^2 + 3x) = \log_5(x + 3)$.

F. Evaluate $\log_{11} 7.13$.

Solution
$$3^{2x} = 5$$
$$\log 3^{2x} = \log 5 \qquad \text{Apply common logs to each side.}$$
$$2x \log 3 = \log 5 \qquad \text{Property of logarithms}$$
$$x = \frac{\log 5}{2 \log 3} = 0.7325$$

To four significant digits, the solution set is {0.7325}.
Now do **A**.

Exercise 14   Solve $x^{-4.1} = 47.6$.

Solution   We take the log of both sides of the equation and proceed as follows: $\log x^{-4.1} = \log 47.6$, so $-4.1 \log x = \log 47.6$, so $\log x = (\log 47.6)/(-4.1)$. To solve for x, evaluate $(\log 47.6)/(-4.1)$ and then find the antilogarithm.

$$x = 47.6 \;\boxed{\text{log}}\; \boxed{\div}\; 4.1 \;\boxed{+/-}\; \boxed{=}\; \boxed{\text{INV}}\; \boxed{\text{log}} \qquad 0.3898$$

Thus, the solution set is { 0.3898}.
Now do **B**.

Exercise 26   Solve $\log(-x) = \frac{1}{2}$.

Solution $\log(-x) = \frac{1}{2}$ is equivalent to $10^{1/2} = -x$, so $x = -\sqrt{10}$, and the solution set is $\{-\sqrt{10}\}$.
Now do **C**.

Exercise 34   Solve $\log_3 x - \log_3 4 = 2$.

Solution $\log_3 x - \log_3 4 = 2$

$$\log_3(x/4) = 2 \qquad \text{Property of logarithms}$$
$$3^2 = x/4, \text{ so } x = 36$$

Thus, the solution set is {36}.
Now do **D**.

Exercise 40   Solve $\log_7(x^2 - x) = \log_7 x + \log_7(x - 1)$.

Solution $\log_7 x + \log_7(x - 1) = \log_7 x(x - 1) = \log_7(x^2 - x)$. Thus, the expressions are identical and the equation is true for all values of x for which the logarithms are defined. Since we can only take the log of positive numbers, the solution set is $(1, \infty)$.
Now do **E**.

Exercise 48   Evaluate $\log_6 0.735$ using common logarithms.

Solution   By the change of base formula, we have

$$\log_6 0.735 = (\log 0.735)/\log 6 = -0.1718.$$

Now do **F**.

Exercise 50   Simplify $\log_x 5 \cdot \log_5 x$.

Solution   Change $\log_x 5$ to base 5 logarithms and proceed as follows:

$$\log_x 5 = \frac{\log_5 5}{\log_5 x} = \frac{1}{\log_5 x}, \text{ so}$$

$$\log_x 5 \cdot \log_5 x = \frac{1}{\log_5 x} \cdot \log_5 x = 1.$$

Now do **G.**

G. Simplify $\log_x 6 \cdot \log_6 x$.

Exercise 52   How long will it take for money invested at 5 percent compounded annually to double?

Solution   If the money doubles at 5 percent compounded annually, then $2P = P(1 + 0.05)^t$ and the equation is $(1.05)^t = 2$. Then

$$t \log 1.05 = \log 2, \text{ so } t = \frac{\log 2}{\log 1.05} = 14.2 \text{ years.}$$

Now do **H.**

H. How long will it take for money invested at 7 percent compounded annually to double?

5.5 More applications and the number e

Objectives checklist

Can you:

a. Simplify exponential and logarithmic statements involving base e?

b. Find the compounded amount for an investment compounded n times per year?

c. Find the compounded amount for an investment compounded continuously?

d. Solve applied problems involving continuous growth or decay by using a base e exponential function?

Key terms

e (irrational number)   Natural logarithm

Key rules and formulas

- *Compound interest formula:* Let A = compounded amount, P = original principal, r = interest rate, n = number of conversions per year, and t = number of years. Then

$$A = P\left(1 + \frac{r}{n}\right)^{nt}.$$

- The number e is an irrational number about equal to 2.71828. . . . This number is the value approached by $\left(1 + \dfrac{1}{n}\right)^n$ as n gets larger and larger.

- Continuous growth or decay: Let A = amount at time t, A_0 = initial amount, k = growth or decay constant. Then if the rate of increase or decrease is proportional to the amount present at any time,

$$A = A_0 e^{kt}.$$

k is positive if the quantity is increasing and negative if the quantity is decreasing.

- The natural logarithm function $y = \log_e x$ is usually written as $y = \ln x$. The functions $y = \ln x$ and $y = e^x$ are inverse functions.

- Using base e, the properties $\log_b b^x = x$ and $b^{\log_b x} = x$ become

$$\ln e^x = x \text{ and } e^{\ln x} = x \ (\text{if } x > 0).$$

5.5

A. If $\ln x = -1.44$, find x.

B. If $e^{3 \ln 5} = b$, find b.

C. If $e^{-2t} = 2.74$, find t.

D. If $y = ce^{t/k}$, find t.

E. You are going to invest $1,000 for 5 years. Which offer is best and what is the compounded amount in this case?
  a. 12 percent compounded monthly
  b. 12.5 percent compounded annually
  c. 11.8 percent compounded continuously

Detailed solutions to selected exercises

Exercise 6 If $\ln x = 2.18$, find x.

Solution $\ln x = 2.18$ is equivalent to $x = e^{2.18}$. By calculator, we have

$$2.18 \;\boxed{\text{INV}}\; \boxed{\text{ln}} \qquad 8.8463.$$

If necessary, we read the less accurate answer 9.0250 from Table 4 in the Appendix.
Now do **A**.

Exercise 10 If $e^{2 \ln 3} = b$, find b.

Solution Although we could use a calculator, it's not necessary. By the laws of logarithms, $e^{2 \ln 3} = e^{\ln 3^2} = 3^2 = 9$.
Now do **B**.

Exercise 16 If $e^{-k} = 0.25$, find k.

Solution
$$
\begin{aligned}
e^{-k} &= 0.25 \\
\ln e^{-k} &= \ln 0.25 && \text{Apply natural logs to each side.} \\
-k &= \ln 0.25 && \text{Use property } \ln e^x = x. \\
k &= 1.3863 && \text{By calculator}
\end{aligned}
$$
Now do **C**.

Exercise 20 If $y = ce^{kt}$, find t.

Solution It's easiest to first divide by c and then apply natural logs to each side as follows:

$$y = ce^{kt}, \text{ so } y/c = e^{kt}, \text{ so } \ln(y/c) = \ln e^{kt} = kt.$$

Then $t = \dfrac{\ln(y/c)}{k}$.
Now do **D**.

Exercise 24 See text for question.

Solution Compute the value of each investment as follows:

a. $A = 10{,}000\left(1 + \dfrac{0.10}{12}\right)^{12(3)} = \$13{,}481.82$

b. $A = 10{,}000(1 + 0.107)^3 = \$13{,}565.72$

c. $A = 10{,}000\, e^{(0.098)(3)} = \$13{,}417.84.$

The second offer is best. (*Note:* Each expression was evaluated by using a calculator.)
Now do **E**.

Exercise 34 A radioactive substance decays from 10 g to 6 g in 5 days. Find the half-life of the substance.

Solution Radioactive substances decay at a rate proportional to the amount present. Thus, the formula is $A = A_0 e^{kt}$. There were 10 g at the start so $A_0 = 10$. Now replace A by 6 and t by 5 and solve for k.

$$6 = 10\, e^{k5}, \text{ so } e^{5k} = 0.6, \text{ so } k = (\ln 0.6)/5 = -0.102$$

To find the half-life, replace A by $\frac{1}{2}A_0$ and solve for t.

$$\tfrac{1}{2}A_0 = A_0 e^{-0.102t}, \text{ so } \ln \tfrac{1}{2} = -0.102t, \text{ so } t = 6.796$$

Thus, the half-life is about 6.8 days.

Now do **F.**

Exercise 36 See text for question.

F. A radioactive substance decays from 100 g to 80 g in 20 days. Find the half-life of the substance.

Solution Substituting the given information into the formula for Newton's law of cooling (see text Example 8), we have

$$T - 40 = (90 - 40)e^{kt}.$$

Now replace T by 60 and t by 1 and solve for k.

$$60 - 40 = 50e^{k \cdot 1}, \text{ so } e^k = 0.4, \text{ so } k = \ln 0.4 = -0.9163.$$

Finally, replace T by 45 and solve for t.

$$45 - 40 = 50e^{-0.9163t}, \text{ so } t = \frac{\ln 0.1}{-0.9163} = 2.5129.$$

Thus, it takes about 2.51 hours.
Now do **G.**

Exercise 42 Show that if $y = \dfrac{r}{1 + ce^{-at}}$, then $t = \dfrac{1}{a} \ln \dfrac{cy}{r - y}$.

G. The temperature of a can of warm soda is 88° F. The soda is placed in a refrigerator with a constant temperature of 43° F. If the soda cools to 58° F in 1 hour, when will the soda's temperature be 46° F?

Solution First, cross multiply and solve for e^{-at} as follows:

$$y + yc(e^{-at}) = r, \text{ so } e^{-at} = (r - y)/cy.$$

Now use natural logarithms and solve for t.

$$-at = \ln \frac{r - y}{cy}, \text{ so } t = -\frac{1}{a} \ln \frac{r - y}{cy} = \frac{1}{a} \ln\left(\frac{r - y}{cy}\right)^{-1} = \frac{1}{a} \ln \frac{cy}{r - y}$$

Now do **H.**

H. If $y = \dfrac{-ce^{at}}{1 - r}$, solve for t.

Sample test questions: Chapter 5

1. Evaluate.

   a. $\log_{10} 10^3$ b. $\ln e$ c. $\log_3 \frac{1}{3}$

   d. $\log_{10} 1$ e. $\log_{10} 123$ f. $\ln 123$

   g. $e^{-0.70}$ h. $e^{2 \ln 5}$ i. $\log_6 9$

2. Complete the statement.

   a. In logarithmic form we write $8^{1/3} = 2$ as _____ .

   b. In exponential form we write $\log_a b = c$ as _____ .

   c. As a single log statement we write $\log(x + h) - \log x$ as _____ .

   d. If $\log 4 = a$ and $\log 5 = b$, then $\log \sqrt{20}$ in terms of a and b is

     _____ .

   e. If $\ln 8 = \ln x - \ln 4$, then x equals _____ .

   f. If the exponential function $y = b^x$ contains the point $(-\frac{1}{3}, 2)$, then b

     equals _____ .

   g. The domain of $y = \ln(x - 3)$ is _____ .

   h. The range of $y = 10^x$ is _____ .

   i. If $f(x) = \log_5 x$, then $f(1)$ equals _____ .

   j. The inverse function of $y = \ln x$ is _____ .

3. Graph

   a. $y = \log_{10} x$ b. $y = (\frac{1}{3})^x$

4. If $g(x) = (\frac{1}{9})^x$, then fill in the missing component in each of the following ordered pairs: $(2, \ \ )$, $(\ , 1)$, $(-1, \ \ )$, $(\ , 81)$, $(\frac{3}{2}, \ \ )$.

5. Solve each equation.

   a. $\log_{10}(3x - 2) = 1$ b. $8^{x-1} = 25$

   c. $\log_x 10 = \frac{1}{2}$ d. $(1.4)^x = 7$

   e. $x^{1.4} = 7$ f. $\log 2 + \log 3 = \log x$

   g. $e^{0.03x} = 0.55$ h. $\log_5 3 = (\ln 3)/\ln x$

   i. $\log_{12}(x + 1) + \log_{12} x = 1$

6. From the following answers, identify *all* choices for k that result in the condition described.

   a. $0 < k < 1$ b. $k > 1$ c. $k = 0$ d. $k = 1$ e. $k < 0$

     (i) $A = A_0 e^{kt}$ describes growth
     (ii) $A = A_0(k)^t$ describes growth
     (iii) $A = A_0 e^{kt}$ describes decay

7. Solve each applied problem.

   a. $\$3,000$ is invested at 8 percent compounded annually. How much is the investment worth in 5 years?

   b. At what rate of interest compounded continuously must money be deposited if the amount is to triple in 10 years?

   c. A radioactive substance decays from 20 g to 5.5 g in 8 days. Find the half-life of the substance.

   d. The Richter scale rating for an earthquake of intensity I equals $\log_{10}(I/I_0)$, where I_0 is a standard reference level number. How many times stronger is a quake rated at 8 than a quake rated at 5?

8. Select the choice that completes the statement or answers the question.

   a. Which function is an exponential function?

     (a) $y = e$ (b) $y = x^e$ (c) $y = x^x$ (d) $y = e^e$ (e) $y = e^x$

   b. $(\log x^n)/(\log y^n)$ simplifies to

     (a) $\dfrac{\log nx}{\log ny}$ (b) $\dfrac{\log x}{\log y}$ (c) $\log x - \log y$

     (d) $\log\left(\dfrac{x}{y}\right)^n$ (e) $\log(x - y)$

   c. If $9^x = \sqrt{27}$, then x equals

     (a) $\frac{1}{3}$ (b) $\frac{3}{2}$ (c) $\frac{3}{4}$ (d) $-\frac{2}{3}$ (e) $\frac{1}{4}$

   d. An expression equivalent to $(\ln 16)/2$ is

     (a) $\ln 2$ (b) $\ln 4$ (c) $\ln 8$ (d) $\ln 14$ (e) $\ln 2$

Chapter 6
The Trigonometric Functions

6.1 Trigonometric functions of acute angles

Objectives checklist

Can you:

a. Find the six trigonometric functions of an acute angle in a right triangle?

b. Find the length of any side in a right triangle given a trigonometric ratio and the length of one side?

c. Find the values of the five remaining trigonometric functions given one function value?

d. Express reciprocal trigonometric functions in terms of each other?

e. Express any trigonometric function of an acute angle as the cofunction of the complementary angle?

f. State exact trigonometric values for 30°, 45°, and 60°?

g. Use a calculator or table to approximate the trigonometric value of a given angle?

h. Use a calculator or table to approximate the measure of an angle given the value of a trigonometric function of that angle?

i. Solve a right triangle?

j. Solve applied problems involving right triangles?

Key terms

Right triangle
Hypotenuse
Acute angle
Complementary angle
Angle of elevation
Angle of depression

Reciprocal function
Cofunctions
Similar triangles
Minute
Significant digits

Key rules and formulas

* *Definition of the trigonometric functions:* For a right triangle,

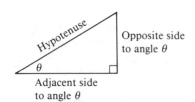

Hypotenuse
Opposite side to angle θ
θ
Adjacent side to angle θ

$$\sin \theta = \frac{\text{opposite}}{\text{hypotenuse}} \quad \text{reciprocals} \quad \csc \theta = \frac{\text{hypotenuse}}{\text{opposite}}$$

$$\cos \theta = \frac{\text{adjacent}}{\text{hypotenuse}} \quad \text{reciprocals} \quad \sec \theta = \frac{\text{hypotenuse}}{\text{adjacent}}$$

$$\tan \theta = \frac{\text{opposite}}{\text{adjacent}} \quad \text{reciprocals} \quad \cot \theta = \frac{\text{adjacent}}{\text{opposite}}$$

Chapter 6

6.1

A. Find the six trigonometric functions of the acute angles in this right triangle.

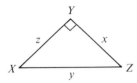

B. In the right triangle shown, represent the ratio 40/41 as the trigonometric function of an acute angle.

C. In the right triangle shown, if $c = 5$ and $a = 3$, find the tangent of the larger acute angle.

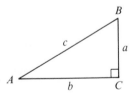

- *Pythagorean theorem:* $c^2 = a^2 + b^2$ (where c is the hypotenuse).

- The sides in a 30-60-90 triangle are related by ratios $1 : \sqrt{3} : 2$.

- The sides in a 45-45-90 triangle are related by ratios $1 : 1 : \sqrt{2}$.

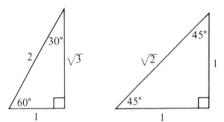

- A trigonometric function of any acute angle is equal to the corresponding cofunction of the complementary angle. The corresponding cofunctions are sine and *co*sine, tangent and *co*tangent, secant and *co*secant.

- To solve a right triangle means to find the measure of the two acute angles and the lengths of the three sides of the triangle. To accomplish this, at least two of these five values must be known and one or more must be the length of a side.

- Results are usually rounded off as follows:

Accuracy of Sides	Accuracy of Angle
Two significant digits	Nearest degree
Three significant digits	Nearest 10 minutes or tenth of a degree
Four significant digits	Nearest minute or hundredth of a degree

Additional comments

- Consult the text for details on evaluating trigonometric expressions with a calculator or table.

- Before attempting the problems in this section, it is suggested that you study the topic of significant digits in the appendix.

- Computed results should not be used to determine the measures of other parts in a triangle when it is possible to use given data.

Detailed solutions to selected exercises

Exercise 4   See text for question.

Solution   First, consider angle R. The length of the opposite side is r, the length of the adjacent side is t, and the length of the hypotenuse is s. Substituting these letters in the given definitions yields

$$\sin R = \frac{\text{opp}}{\text{hyp}} = \frac{r}{s} \qquad \text{reciprocals} \qquad \csc R = \frac{\text{hyp}}{\text{opp}} = \frac{s}{r}$$

$$\cos R = \frac{\text{adj}}{\text{hyp}} = \frac{t}{s} \qquad \text{reciprocals} \qquad \sec R = \frac{\text{hyp}}{\text{adj}} = \frac{s}{t}$$

$$\tan R = \frac{\text{opp}}{\text{adj}} = \frac{r}{t} \qquad \text{reciprocals} \qquad \cot R = \frac{\text{adj}}{\text{opp}} = \frac{t}{r}.$$

The trigonometric functions of T may be found in a similar way. However, it is more efficient to recognize that R and T are complementary angles and that cofunctions of complementary angles are equal. Thus, $\sin R = \cos T = r/s$, $\cos R = \sin T = t/s$, $\tan R = \cot T = r/t$, $\cot R = \tan T = t/r$, $\sec R = \csc T = s/t$, and $\csc R = \sec T = s/r$.
Now do **A**.

Exercise 12 See text for question.

Solution With respect to angle θ, $\frac{9}{40}$ corresponds to $\frac{\text{adj}}{\text{opp}}$, which is $\cot \theta$. In terms of angle α, we determine $\frac{9}{40} = \tan \alpha$ by either using the concept of cofunctions or recognizing that $\frac{9}{40}$ corresponds to $\frac{\text{opp}}{\text{adj}}$. Thus, $\frac{9}{40} = \cot \theta = \tan \alpha$.
Now do **B**.

Exercise 18 If $c = 17$ and $a = 8$, find the tangent of the larger acute angle.

Solution From the Pythagorean theorem $c^2 = a^2 + b^2$, we have $17^2 = 8^2 + b^2$, so $b = 15$. Since $b > a$, angle B is the larger acute angle. Then $\tan B = b/a = \frac{15}{8}$.
Now do **C**.

Exercise 24 If $\sec \theta = 5$, find values for the other trigonometric functions.

Solution If $\sec \theta = 5$, then $\cos \theta = \frac{1}{5}$. To find the four remaining functions, since $\sec \theta = 5$, we can let the length of the hypotenuse be 5 and the length of the adjacent side be 1. By the Pythagorean theorem, we then determine that the length of the opposite side is $\sqrt{24}$. Finally, using the function definitions, we have $\sin \theta = \sqrt{24}/5$, $\csc \theta = 5/\sqrt{24}$, $\tan \theta = \sqrt{24}$, and $\cot \theta = 1/\sqrt{24}$.
Now do **D**.

Exercise 30 Express $\cot 0.5°$ as a function of the angle complementary to the given angle.

Solution Since cofunctions of complementary angles are equal, $\cot 0.5° = \tan(90° - 0.5°) = \tan 89.5°$.
Now do **E**.

Exercise 32 Does $\sin 60° \cdot \csc 30° = \tan 60°$?

Solution The exact values are $\sin 60° = \sqrt{3}/2$, $\csc 30° = 2$, and $\tan 60° = \sqrt{3}$. Since $(\sqrt{3}/2) \cdot 2 = \sqrt{3}$, the answer is "yes."
Now do **F**.

Exercise 42 Approximate $\csc 54°$.

Solution Calculator: First, make sure the calculator is in degree mode. Then since cosecant and sine are reciprocals, use

$$54 \;\boxed{\sin}\; \boxed{1/x} \qquad 1.236068.$$

Table: Since the angle is greater than $45°$, the angle is listed in the right-hand column and the function name is read from the bottom row. We read $\csc 54° \approx 1.236$.
Now do **G**.

Exercise 58 Approximate θ if $\sec \theta \approx 1.549$.

Solution Calculator: Since there is no button for secant, we must recognize that $\sec \theta = 1.549$ is equivalent to $\cos \theta = \frac{1}{1.549}$. Then

$$1.549 \;\boxed{1/x}\; \boxed{\text{INV}}\; \boxed{\cos} \qquad 49.79099.$$

D. If $\csc \theta = \sqrt{2}$, find values for the other trigonometric functions.

E. Express $\csc 30.5°$ as a function of the angle complementary to the given angle.

F. Does
$$\cos 60° \cdot \sin 30° = \cot 60°?$$

G. Approximate $\sec 36°$.

H. Approximate θ if $\csc \theta = 1.467$.

Table: The function name is read from the bottom and the angle is listed in the right-hand column. The closest entry to 1.549 is 1.550, so $\csc 49°50' \approx 1.549$.

Now do **H.**

Exercise 78   Solve right triangle ABC ($C = 90°$) if $a = 7.00$ ft and $c = 11.0$ ft.

Solution   First, sketch Figure 6.1:78. Now find b by the Pythagorean theorem ($c^2 = a^2 + b^2$).

$$(11.0)^2 = (7.00)^2 + b^2 \quad \text{so} \quad b = \sqrt{72} = 8.49 \text{ ft}$$

I. Solve right triangle ABC ($C = 90°$) if $a = 6.00$ ft and $c = 9.00$ ft.

Second, we find angle A by using the sine function.

$$\sin A = a/c = \tfrac{7}{11} = .63636$$
$$A = 39°30' \text{ or } 39.5°$$

Third, since $A + B = 90°$, $B = 50°30'$ or $50.5°$.

J. If a 15-foot ladder rests against a 40-foot building and the angle of elevation is 32°, how far from the base of the building is the base of the ladder?

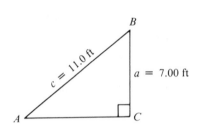

Figure 6.1:78   **Figure 6.1:84**

Now do **I.**

Exercise 84   See text for question.

Solution   First, sketch Figure 6.1:84. The ceiling (x) is the side opposite 58° and we know the adjacent side length. Thus,

$$\tan 58° = x/600 \quad \text{so} \quad x = 600 \tan 58° = 960 \text{ ft.}$$

Now do **J.**

Exercise 90   See text for question.

Solution   First, sketch Figure 6.1:90. The length of the hypotenuse is x while the length of the side adjacent to θ is $x/4$. Thus,

$$\cos \theta = \frac{x/4}{x} = \frac{1}{4} = .25$$
$$\theta = 75°30' \text{ or } 75.5°.$$

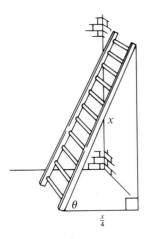

Figure 6.1:90

Now do **K.**

K. A ladder is set up so that the distance between the base of the ladder and the building against which it leans is one-half the length of the ladder. Find the angle the ladder makes with the ground.

6.2 Trigonometric functions of general angles

Objectives checklist

Can you:

a. Find the values of the six trigonometric functions of an angle given the terminal ray of the angle?

b. Find the values of the five remaining trigonometric functions of θ given one function value and the quadrant containing the terminal ray of θ?

c. Find the reference angle for a given angle?

d. Determine the exact value of any trigonometric function of quadrantal angles or angles with reference angles of 30°, 45°, or 60°?

e. Determine the approximate trigonometric value for any angle using a calculator or a table and reference angles?

Key terms

Ray   Measure of an angle
Initial ray   Quadrantal angle
Terminal ray   Reference angle
Standard position of θ Coterminal angle

Key rules and formulas

• *Definition of the trigonometric function:* For a general angle,

$$\sin \theta = \frac{y}{r} \qquad \text{reciprocals} \qquad \csc \theta = \frac{r}{y}$$

$$\cos \theta = \frac{x}{r} \qquad \text{reciprocals} \qquad \sec \theta = \frac{r}{x}$$

$$\tan \theta = \frac{y}{x} \qquad \text{reciprocals} \qquad \cot \theta = \frac{x}{y}$$

6.2

A. Find the values of the six trigonometric functions of θ if the terminal ray of θ lies in Q_3 on the line $y = 4x$.

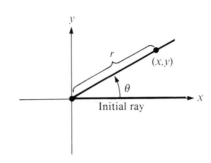

Note: $r = \sqrt{x^2 + y^2}$ or $r^2 = x^2 + y^2$; x and y cannot be 0 in the denominator of a ratio.

- If two angles have the same terminal ray, they are called coterminal and the trigonometric functions of coterminal angles are equal.

B. Find the values of the remaining trigonometric functions of θ if $\sec \theta = 2$ and $\sin \theta > 0$.

- *Signs of the trigonometric ratios:*

$$
\begin{array}{c|c}
\left.\begin{array}{c}\sin \theta \\ \csc \theta\end{array}\right\} + & \underline{\text{All the functions}} \\
& \text{are positive} \\
\text{others} \} - & \\
\hline
\left.\begin{array}{c}\tan \theta \\ \cot \theta\end{array}\right\} + & \left.\begin{array}{c}\cos \theta \\ \sec \theta\end{array}\right\} + \\
\text{others} \} - & \text{others} \} -
\end{array}
$$

C. Find the reference angle for $950°$.

A good memory aid is the sentence "*A*ll *s*tudents *t*ake *c*alculus."

- *To evaluate trigonometric functions (for nonquadrantal angles):*

  1. Find the reference angle for the given angle.
  2. Find the trigonometric value of the reference angle using the appropriate function and Table 5. (*Note:* If the reference angle is 30°, 45°, or 60°, the exact answer is preferable.)
  3. Determine the correct sign according to the terminal ray of the angle and the chart showing the signs of the trigonometric ratios.

Additional comments

- Right triangle trigonometry is a special case of the more general definition of the trigonometric functions for it arises when we are dealing with an acute angle whose terminal ray lies in the first quadrant.

- The back endpaper of the text contains a chart with the exact trigonometric values of 0°, 30°, 45°, 60°, 90°, 180°, and 270°.

Detailed solutions to selected exercises

Exercise 10 Find the values of the six trigonometric functions of θ if the terminal ray of θ lies in Q_4 on the line $y = -2x$.

Solution Pick any point in Q_4 satisfying $y = -2x$, say, $x = 1$, $y = -2$. Then $r = \sqrt{x^2 + y^2} = \sqrt{1^2 + (-2)^2} = \sqrt{1 + 4} = \sqrt{5}$. By definition,

$$
\begin{array}{lll}
\sin \theta = y/r = -2/\sqrt{5} & \text{reciprocals} & \csc \theta = -\sqrt{5}/2 \\
\cos \theta = x/r = 1/\sqrt{5} & \text{reciprocals} & \sec \theta = \sqrt{5} \\
\tan \theta = y/x = -2 & \text{reciprocals} & \cot \theta = -\tfrac{1}{2}.
\end{array}
$$

Now do **A**.

Exercise 24 Find the values of the remaining trigonometric functions of θ if $\csc \theta = 2$ and $\cos \theta < 0$.

Solution If $\csc \theta = 2$, then $\sin \theta = \frac{1}{2}$. To find the values for the four remaining functions, we note that $\csc \theta$ is positive in Q_1 and Q_2, while $\cos \theta$ is negative in Q_2 and Q_3. Only Q_2 satisfies both conditions so θ is in the second quadrant, where x is negative and y is positive. Since $\csc \theta = r/y = 2$, we can set $r = 2$ and $y = 1$ and find x by using $r^2 = x^2 + y^2$.

$$2^2 = x^2 + 1^2 \quad \text{so} \quad x = \pm \sqrt{3}$$

Because θ is in Q_2, we conclude $x = -\sqrt{3}$. Then by definition, $\cos \theta = -\sqrt{3}/2$, $\sec \theta = -2/\sqrt{3}$, $\tan \theta = -1/\sqrt{3}$, and $\cot \theta = -\sqrt{3}$.
Now do **B**.

Exercise 32 Find the reference angle for $1{,}000°$.

Solution The reference angle for $1{,}000°$ is $80°$ since the closest segment of the horizontal axis (positive x-axis) may correspond to a rotation of $3 \cdot 360°$ or $1{,}080°$ and $|1{,}080° - 1{,}000°| = 80°$.
Now do **C**.

Exercise 44 Find the exact value of $\sin 630°$.

Solution $630°$ is a quadrantal angle that is coterminal with $270°$. Then by the chart on the back endpaper of the text, $\sin 630° = \sin 270° = -1$.
Now do **D**.

Exercise 56 Find the exact value of $\sec(-225°)$.

Solution The negative x-axis may correspond to a rotation of $-180°$ so the reference angle is $|-180° - (-225°)| = 45°$. Then $\sec 45° = \sqrt{2}$ from either the chart on the back endpaper or the $1{:}1{:}\sqrt{2}$ ratios in a 45-45-90 triangle. Finally, a rotation of $225°$ in a clockwise direction produces an angle in Q_2, where $\sec \theta$ is negative. Thus, $\sec(-225°) = -\sqrt{2}$.
Now do **E**.

Exercise 68 Find the approximate value of $\cot 115°$.

Solution The reference angle is $|180° - 115°| = 65°$. From Table 5, $\cot 65° = 0.4663$. Finally, $115°$ is in Q_2, where $\cot \theta < 0$. Thus, $\cot 115° = -\cot 65° = -0.4663$. By calculator,

$$115 \boxed{\tan} \boxed{1/x} \qquad -0.4663076.$$

Now do **F**.

6.3 Radians

Objectives checklist

Can you:

a. Convert from degree measure of an angle to radian measure, and vice versa?

b. Use $\theta = s/r$ to find either the central angle, the intercepted arc, or the radius given measures for two variables in the relationship?

c. Determine the area of a sector of a circle given the radius and the central angle?

d. Find linear velocity given an angular velocity and a radius?

e. Find angular velocity given a linear velocity and a radius?

D. Find the exact value of $\sin 1{,}170°$.

E. Find the exact value of $\csc(-225°)$.

F. Find the approximate value of $\tan 115°$.

6.3

A. Find the length of the arc intercepted by a central angle of 4.1 radians in a circle whose radius is 9 feet.

B. In a circle of radius 8 ft, the arc length of a sector is 5 ft. Find the area of the sector.

Key terms

Radian   Angular velocity
Linear velocity

Key rules and formulas

- In a circle of radius r, let s be the length of the arc intercepted by central angle θ. Then

$$\theta = s/r$$

is the radian measure of the angle. Equivalently, an angle measuring 1 radian intercepts an arc equal in length to the radius of the circle. We can find the radian measure of θ by using a circle of any radius.

- Since $360° = 2\pi$ radians, $1° = \pi/180$ radians, and 1 radian $= 180°/\pi$. Then

  1. To change degree measure to radian measure, multiply the degree measure by $\pi/180°$ (and simplify if possible).
  2. To change radian measure to degree measure, multiply the radian measure by $180°/\pi$ (and simplify if possible).

- The area A of the sector of a circle with radius r and central angle θ radians is given by

$$A = \tfrac{1}{2}r^2\theta.$$

- If a point P moves along the circumference of a circle at a constant speed, then the linear velocity v is equal to the product of the angular velocity ω (omega) and the radius. In symbols,

$$v = \omega r.$$

Additional comments

- Remember that θ must be expressed in radians to use $s = \theta \cdot r$ and $A = \tfrac{1}{2}r^2\theta$, while ω must be expressed in radians per unit of time to use $v = \omega r$.

Detailed solutions to selected exercises

Exercise 6 Find the length of the arc intercepted by a central angle of 3.2 radians in a circle whose radius is 11 ft.

Solution Substituting $r = 11$ ft and $\theta = 3.2$ in the formula $\theta = s/r$, we have $3.2 = s/11$ ft so $s = 35.2$ ft. For such cases, the formula $\theta = s/r$ is usually written as $s = \theta \cdot r$.
Now do **A.**

Exercise 14 In a circle of radius 5 ft, the arc length of a sector is 3 ft. Find the area of the sector.

Solution First, the central angle $\theta = s/r = 3$ ft/5 ft $= \tfrac{3}{5}$. Substituting $\theta = \tfrac{3}{5}$ and $r = 5$ ft in the formula $A = \tfrac{1}{2}r^2\theta$, we have

$$A = \tfrac{1}{2}(5 \text{ ft})^2 \cdot \tfrac{3}{5} = \tfrac{15}{2} \text{ ft}^2.$$

Now do **B.**

Exercise 24 Express 225° in radian measure.

Solution Using the conversion factor $\pi/180°$, we have

$$225° = 225 \cdot \frac{\pi}{180} = \frac{225\pi}{180} = \frac{5\pi}{4}.$$

Now do **C**.

Exercise 44 Express $2\pi/15$ radians in degree measure.

Solution Using the conversion factor $180°/\pi$, we have

$$\frac{2\pi}{15} = \frac{2\pi}{15} \cdot \frac{180°}{\pi} = \frac{360°}{15} = 24°.$$

Now do **D**.

Exercise 52 To the nearest degree, express 3.5 radians in degree measure.

Solution Using the conversion factor $180°/\pi$, we have

$$3.5 = 3.5\frac{180°}{\pi} = \frac{630°}{\pi} = 200.535°.$$

To the nearest degree, the answer is 201°.
Now do **E**.

Exercise 60 A car is traveling at 60 mi/hour (88 ft/second) on tires 30 in. in diameter. Find the angular velocity of the tires in radians/sec.

Solution If the diameter is 30 in., the radius is 15 in. or $\frac{5}{4}$ ft. Substituting this value for r and 88 ft/second for the linear velocity (v) in the formula $v = \omega r$ gives

$$88\,\frac{\text{ft}}{\text{sec}} = \omega\left(\frac{5}{4}\,\text{ft}\right) \text{ so } \omega = 88\,\frac{\text{ft}}{\text{sec}} \div \frac{5}{4}\,\text{ft} = \frac{352}{5}\,\frac{\text{radians}}{\text{second}}.$$

Now do **F**.

Exercise 62 See text for question.

Solution

a. Consider one loop in the quatrefoil as shown in Figure 6.3:62. The arc length from B to C is given by $s = \theta \cdot r = (\pi/2) \cdot 2 = \pi$. There are two such segments in the figure and eight such segments in the quatrefoil. Thus, the perimeter is 8π inches.

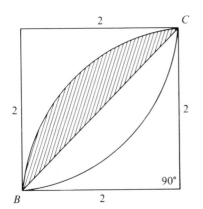

Figure 6.3:62

C. Express 345° in radian measure.

D. Express $5\pi/12$ radians in degree measure.

E. To the nearest degree, express 7.21 radians in degree measure.

F. A car is traveling at 30 mi/hr (44 ft/sec) on tires 28 in. in diameter. Find the angular velocity of the tires in radians/sec.

G. Consider the quatrefoil in the figure. If the sides of the square measure 10 in., determine
  a. the perimeter of the quatrefoil
  b. the area of the quatrefoil

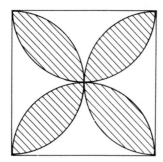

b. The area of the shaded region in Figure 6.3:62 is given by the area of the sector of the circle minus the area of the triangle. Thus,

$$A = \tfrac{1}{2}r^2\theta - \tfrac{1}{2}bh$$
$$= \tfrac{1}{2}(2)^2 \cdot \pi/2 - \tfrac{1}{2}(2)(2)$$
$$= \pi - 2.$$

The quatrefoil has 8 such regions, so the total area is $8\pi - 16$ in². Now do **G.**

6.4 Trigonometric functions of real numbers

Objectives checklist

Can you:

a. Find the reference angle (or number) for a given angle (or number)?

b. Determine the exact value of any trigonometric function of quadrantal angles or angles with reference angles of $\pi/6$, $\pi/4$, or $\pi/3$?

c. Determine the approximate trigonometric value for any angle using a calculator or a table and reference angles?

Key terms

Unit circle   Reference angle (or number)
Trigonometric identity

Key rules and formulas

*   *Definitions of the trigonometric functions:* For a unit circle,

$$\sin s = y \qquad\qquad \text{reciprocals} \qquad \csc s = 1/y = 1/\sin s$$
$$\cos s = x \qquad\qquad \text{reciprocals} \qquad \sec s = 1/x = 1/\cos s$$
$$\tan s = y/x = \frac{\sin s}{\cos s} \quad \text{reciprocals} \quad \cot s = x/y = \frac{\cos s}{\sin s}$$

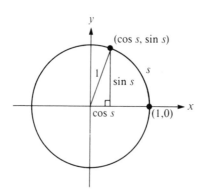

*   For the sine and cosine functions, the domain is the set of all real numbers and the range is the set of real numbers between -1 and 1, inclusive.

*   For any trigonometric function f, we have $f(s + 2\pi k) = f(s)$, where k is an integer.

- *Negative angle (or number) identities:*

$$\sin(-s) = -\sin s \qquad \csc(-s) = -\csc s$$
$$\cos(-s) = \cos s \qquad \sec(-s) = \sec s$$
$$\tan(-s) = -\tan s \qquad \cot(-s) = -\cot s$$

6.4

A. Find the exact value of $\cos(-7\pi/4)$.

- To find (without a calculator) the trigonometric value of any angle (or number), do the following:

  1. If necessary, use $f(s + 2\pi k) = f(s)$ and the negative angle identities to find an equivalent expression such that $0 \le s < 2\pi$.
  2. If the angle terminates at one of the axes, use the function definition and the appropriate x- and y-coordinates to find the function value. Otherwise, we need to find the reference angle for s by finding the shortest positive arc length between the point on the circle assigned to s and the x-axis.
  3. Find the trigonometric value of the reference angle using the appropriate function and Table 6. (*Note:* If the reference angle is $\pi/6$, $\pi/4$, or $\pi/3$, the exact answer is preferable.)
  4. Determine the correct sign by using the following chart.

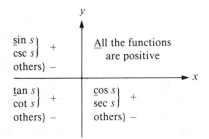

Additional comments

- The back endpaper of the text contains a chart with the exact trigonometric values of 0, $\pi/6$, $\pi/4$, $\pi/3$, $\pi/2$, π, and $3\pi/2$.

- To evaluate trigonometric functions of real numbers, be sure to set the calculator for radian mode. Most calculators are in degree mode when first turned on, and forgetting this step is a very common error. For other notes on calculator usage, consult the text.

Detailed solutions to selected exercises

Exercise 12 Find the exact value of $\cos(-9\pi/2)$.

Solution Since $\cos(-s) = \cos s$, we have $\cos(-9\pi/2) = \cos(9\pi/2)$. Then

$$\cos\left(\frac{9\pi}{2}\right) = \cos\left(\frac{\pi}{2} + 4\pi\right) = \cos\frac{\pi}{2}.$$

Finally, the x-coordinate on the unit circle when $s = \pi/2$ is 0. Thus, $\cos(-9\pi/2) = 0$. (*Note:* $\cos(\pi/2) = 0$ may also be read from the chart on the back endpaper of the text.)

Calculator solution 9 $\boxed{+/-}$ $\boxed{\times}$ π $\boxed{\div}$ 2 $\boxed{=}$ $\boxed{\cos}$ 0
(*Note:* If you keep getting 0.9697 as the answer on your calculator, the problem is that the machine is incorrectly set in degree mode.)
Now do **A.**

B. Find the exact value of sec($13\pi/6$).

C. Find the exact value of csc($29\pi/3$).

D. Approximate sin(-11.83).

Exercise 20 Find the exact value of sec($17\pi/4$).

Solution sec($17\pi/4$) = sec$[(\pi/4) + 4\pi]$ = sec($\pi/4$). Then sec($\pi/4$) equals $\sqrt{2}$ from either the chart on the back endpaper of the text or the $1:1:\sqrt{2}$ ratios in a 45-45-90 triangle. Thus, sec($17\pi/4$) = $\sqrt{2}$.

Calculator solution 17 $\boxed{\times}$ π $\boxed{\div}$ 4 $\boxed{=}$ $\boxed{\cos}$ $\boxed{1/x}$ 1.4142
(*Note:* The calculator solution is not an exact value as required in the question.)
Now do **B.**

Exercise 48 Find the exact value of csc($15\pi/4$).

Solution We follow the steps outlined above.

1. $\csc\left(\dfrac{15\pi}{4}\right) = \csc\left(\dfrac{7\pi}{4} + 2\pi\right) = \csc\dfrac{7\pi}{4}$.
2. Since $2\pi - 7\pi/4 = \pi/4$, the reference angle is $\pi/4$.
3. csc $\pi/4 = \sqrt{2}$ (using chart or 45-45-90 triangle).
4. The point on the unit circle assigned to $7\pi/4$ is in Q_4, where the value of the cosecant function is negative.

Thus, csc($15\pi/4$) = $-\sqrt{2}$.
Now do **C.**

Exercise 64 Approximate sin(-10.25).

Solution We follow the steps outlined above.

1. $\sin(-10.25) = -\sin 10.25 = -\sin(3.97 + 2\pi) = -\sin 3.97$.
2. Since $3.97 - \pi = 3.97 - 3.14 = 0.83$, the reference number is 0.83.
3. $\sin 0.83 = 0.7379$ (using Table 6 in the text).
4. The point on the unit circle assigned to 3.97 is in Q_3, where the value of the sine function is negative.

Thus, $\sin(-10.25) = -(-\sin 0.83) = 0.7379$.

Calculator solution 10.25 $\boxed{+/-}$ sin 0.7347
(*Note:* The calculator solution is more accurate in approximation problems like this one.)
Now do **D.**

6.5 Graphs of sine and cosine functions

Objectives checklist

Can you:

a. Find the amplitude and period for functions of the form $y = a \sin bx$ and $y = a \cos bx$ and graph the function for a given interval?

b. Write the equation of the form $y = a \sin bx$ or $y = a \cos bx$ which corresponds to a given curve?

c. Find the amplitude, period, and phase shift for functions of the form $y = a \sin(bx + c)$ and $y = a \cos(bx + c)$ and graph the function for a given interval?

Key terms

Periodic function   Amplitude
Period   Phase shift

Key rules and formulas

- *Period:* A function f is periodic if $f(x) = f(x + p)$ for all x in the domain of f. The smallest positive number p for which this is true is called the period of the function. The sine and cosine functions are periodic with period 2π. For $y = a \sin(bx + c)$ and $y = a \cos(bx + c)$, we determine the period as follows:

$$\text{Period} = 2\pi/b.$$

- *Amplitude:* When the graph of a periodic function is centered about the x-axis, the maximum y value is called the amplitude of the function. The amplitude for the sine and cosine functions is 1. For $y = a \sin(bx + c)$ and $y = a \cos(bx + c)$, we determine the amplitude as follows:

$$\text{Amplitude} = |a|.$$

- *Phase shift:* The constant c in the function $y = a \sin(bx + c)$ causes a shift of the graph of $y = a \sin bx$. The shift is of distance $|c/b|$ and is to the left if $c > 0$ and to the right if $c < 0$. We can compute the phase shift as follows:

$$\text{Phase shift} = -c/b.$$

Similar remarks hold for functions of the form $y = a \cos(bx + c)$.

- The basic cycle for $y = \sin x$ starts at the origin, attains a maximum at $x = \pi/2$ (one-fourth of the cycle length), returns to zero at $x = \pi$ (halfway through the cycle), attains a minimum at $x = 3\pi/2$ (the three-quarter point), and returns to zero at $x = 2\pi$ (the end of the cycle). The cosine function oscillates in a similar way except that $y = \cos x$ starts its basic cycle by attaining its maximum height at $x = 0$.

A. Graph $y = 2 \sin 3x$
 ($-\text{period} \le x \le \text{period}$).

Detailed solutions to selected exercises

Exercise 8   Graph $y = -6 \sin(x/4)$ ($-\text{period} \le x \le \text{period}$).

Solution   Amplitude $= |a| = |-6| = 6$

$$\text{Period} = \frac{2\pi}{b} = \frac{2\pi}{1/4} = 8\pi \ \left(\textit{Note:} \ \frac{x}{4} = \frac{1}{4}x\right)$$

Since the amplitude is 6, the curve oscillates between a maximum value of 6 and a minimum value of -6. Because a is negative, we proceed from the origin to the minimum value of -6 at $x = 2\pi$. For $-8\pi \le x \le 8\pi$, we draw two cycles in the graph as shown in Figure 6.5:8.

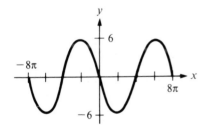

Figure 6.5:8

Now do **A.**

B. Graph $y = -2 \cos(\pi/3)x$
$(0 \le x \le 2\pi)$.

C. Graph one cycle of
$y = -\cos(2x + \pi/2)$.

Exercise 18 Graph $y = 2 \cos \pi x \ (0 \le x \le 2\pi)$.

Solution Amplitude $= |a| = |2| = 2$

$$\text{Period} = \frac{2\pi}{b} = \frac{2\pi}{\pi} = 2$$

If the curve completes one cycle on the interval [0,2], then the curve completes a little more than 3 cycles on the interval $[0,2\pi]$. Since the amplitude is 2, the curve oscillates between a maximum value of 2 and a minimum value of -2. At $x = 0$, the cosine function starts at the maximum function value. The graph is shown in Figure 6.5:18.

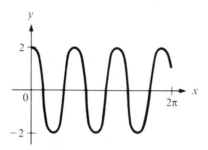

Figure 6.5:18

Now do **B.**

Exercise 30 Graph one cycle of $y = 1.2 \cos\left(2\pi x - \frac{\pi}{2}\right)$.

Solution Amplitude $= |a| = |1.2| = 1.2$

$$\text{Period} = \frac{2\pi}{b} = \frac{2\pi}{2\pi} = 1$$

$$\text{Phase shift} = \frac{-c}{b} = \frac{-(-\pi/2)}{2\pi} = \frac{1}{4}$$

One cycle starts at $\frac{1}{4}$ and ends at $\frac{5}{4}$, which is $\frac{1}{4}$ plus the period. The graph is shown in Figure 6.5:30.

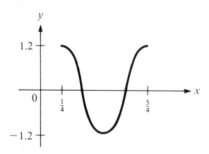

Figure 6.5:30

Now do **C.**

Exercise 38 See text for question.

Solution Since the cycle starts at the minimum y value, the form of the equation is $y = a \cos bx$ with $a < 0$. The amplitude is 2.5 and $a < 0$, so $a = -2.5$. Since the period is $3\pi/2$, we have

$$\frac{3\pi}{2} = \frac{2\pi}{b} \quad \text{so} \quad b = \frac{4\pi}{3\pi} = \frac{4}{3}.$$

Thus, an equation for the graph is $y = -2.5 \cos \frac{4}{3}x$.
Now do **D**.

Exercise 46 See text for question.

Solution The period for $y = 2 \sin 3x$ is $2\pi/3$. The sine function attains its maximum value at one-fourth of the cycle length, so $y = 2 \sin 3x$ reaches its maximum value when $x = \frac{1}{4}\left(\frac{2\pi}{3}\right) = \frac{\pi}{6}$ (choice b).

Alternate method: The maximum function value results for $\sin(\pi/2)$. Setting $3x$ equal to $\pi/2$ gives $x = \pi/6$.
Now do **E**.

Exercise 60 See text for question.

Solution Use the definition given in the question as follows:

a. $M = 1, m = -1$, amplitude $= \dfrac{1 - (-1)}{2} = 1$

b. $M = |a|, m = -|a|$, amplitude $= \dfrac{|a| - (-|a|)}{2} = \dfrac{2|a|}{2} = |a|$

c. $M = 4, m = 2$, amplitude $= \dfrac{4 - 2}{2} = 1$

d. $M = 1, m = 0$, amplitude $= \dfrac{1 - 0}{2} = \dfrac{1}{2}$, Period: 2

Now do **F**.

6.6 Graphs of the other trigonometric functions

Objectives checklist

Can you:

a. Graph tangent, cotangent, secant, and cosecant functions?

b. State the domain, range, and period for $y = \tan x$, $y = \cot x$, $y = \sec x$, and $y = \csc x$?

c. Determine for each trigonometric function whether it is increasing or decreasing on a given interval?

Key rules and formulas

• From the graph of the tangent function, we note the following features:

Domain: Set of all real numbers except $x = (\pi/2) + k\pi$ (k any integer)
Range: Set of all real numbers
Period: π

D. Find an equation of the form $y = a \sin bx$ or $y = a \cos bx$ for this curve. A single cycle is shown.

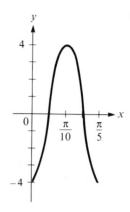

E. For which value of x does the expression $y = 4 \sin \frac{1}{3}x$ reach its maximum value? (Choose one)
a. $3\pi/2$
b. $\pi/6$
c. $\pi/2$
d. 2π

F. Using the definition of amplitude (amplitude $= (M - m)/2$) (see Exercise 60),
a. show the amplitude of $y = \cos x$ is 1.
b. show the amplitude of $y = a \cos x$ is $|a|$.
c. show the amplitude of $y = 2 + \cos x$ is 1.

6.6

A. Complete the table shown in the text for $y = \csc x$ and then sketch the curve from these points on the interval $[0,2\pi]$.

B. Find the domain, range, and period for $y = 2 \csc x$.

C. Use the graph of $y = -\cot x$ to determine whether the function is increasing or decreasing in the intervals $(0,\pi/2)$, $(\pi/2,\pi)$, $(\pi,3\pi/2)$, and $(3\pi/2,2\pi)$.

• We graph $y = \csc x$, $y = \sec x$, and $y = \cot x$ as the reciprocals of the sine, cosine, and tangent functions, respectively. Note the following about a function and its reciprocal function.

1. As one function increases, the other decreases, and vice versa.
2. The two functions always have the same sign.
3. When one function is zero, the other is undefined.
4. When the value of the function is 1 or -1, the reciprocal function has the same value.

Detailed solutions to selected exercises

Exercise 2   See text for question.

Solution   The completed table is as follows:

x	0	$\dfrac{\pi}{6}$	$\dfrac{\pi}{3}$	$\dfrac{\pi}{2}$	$\dfrac{2\pi}{3}$	$\dfrac{5\pi}{6}$	π	$\dfrac{7\pi}{6}$	$\dfrac{4\pi}{3}$	$\dfrac{3\pi}{2}$	$\dfrac{5\pi}{3}$	$\dfrac{11\pi}{6}$	2π
$\sec x$	1	1.2	2	und.	-2	-1.2	-1	-1.2	-2	und.	2	1.2	1

These values may be obtained by calculator or by using the chart on the back endpaper of the text together with the concept of reference numbers. The graph for $y = \sec x$ on $[0,2\pi]$ is given in Figure 6.6:2.

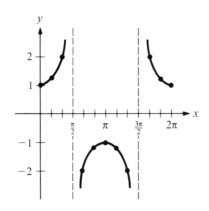

Figure 6.6:2

Now do **A.**

Exercise 6   Find the domain, range, and period for $y = \csc x$.

Solution   Noting that $\csc x = 1/\sin x$ and considering the graph of $y = \csc x$ in Figure 6.68 in the text, we conclude the following.

Domain: Set of all real numbers except $x = k\pi$ (k is any integer)
Range: $(-\infty,-1] \cup [1,\infty)$
Period: 2π

Now do **B.**

Exercise 10   See text for question.

Solution   As x increases in each of the given intervals, $\cot x$ keeps getting smaller. Thus, the cotangent function is a decreasing function in all 4 intervals. Now do **C.**

Exercise 18   Graph one cycle of $y = \sec 2x$.

Solution First, sketch $y = \cos 2x$ as shown by the dotted lines in Figure 6.6:18. Note the amplitude of $y = \cos 2x$ is 1, while the period is π. Then $y = \sec 2x = 1/\cos 2x$, so we graph the function as shown by obtaining the reciprocals of the various y values of the cosine function.

D. Graph one cycle of $y = \sec 2\pi x$.

E. Graph one cycle of $y = 3 \csc(x/2)$.

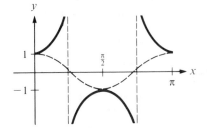

Figure 6.6:18

Now do **D.**

Exercise 20 Graph one cycle of $y = 2 \csc \pi x$.

Solution Since $\csc \pi x = 1/\sin \pi x$, we first sketch $y = 2 \sin \pi x$ to use as a reference. The amplitude and period for $y = 2 \sin \pi x$ are both 2 and the dotted lines in Figure 6.6:20 indicate the graph for this function. When $\sin \pi x = 0$, $\csc \pi x$ is undefined with the cosecant graph having vertical asymptotes at these x values (which are $x = 0$, $x = 1$, and $x = 2$). When $2 \sin \pi x$ attains its maximum value at $x = \frac{1}{2}$, $2 \csc \pi x$ attains a minimum turn value at $(\frac{1}{2}, 2)$. Similarly, when $2 \sin \pi x$ attains its minimum value at $x = \frac{3}{2}$, $2 \csc \pi x$ attains a maximum turning point at $(\frac{3}{2}, -2)$. Using these key features, we then complete the graph as shown in Figure 6.6:20.

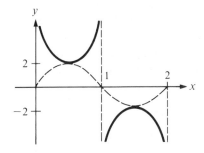

Figure 6.6:20

Now do **E.**

6.7 Trigonometric identities

Objectives checklist

Can you:

a. State the eight fundamental identities listed in this section?

b. Transform a given trigonometric expression and show that it is identical to another given expression?

c. Verify that certain trigonometric equations are identities?

6.7

A. Show that $\cos x(\sec x - \cos x)$ is identical to $\sin^2 x$.

Key Terms

Trigonometric identity   Pythagorean identities
Fundamental identities

Key rules and formulas

- A trigonometric identity is a statement that is true for all values of x for which the expressions are defined. The eight fundamental identities are as follows:

  Identity 1 $\csc x = \dfrac{1}{\sin x}$ or $\sin x = \dfrac{1}{\csc x}$ or $\sin x \csc x = 1$

  Identity 2 $\sec x = \dfrac{1}{\cos x}$ or $\cos x = \dfrac{1}{\sec x}$ or $\cos x \sec x = 1$

  Identity 3 $\cot x = \dfrac{1}{\tan x}$ or $\tan x = \dfrac{1}{\cot x}$ or $\tan x \cot x = 1$

  Identity 4 $\tan x = \dfrac{\sin x}{\cos x}$

  Identity 5 $\cot x = \dfrac{\cos x}{\sin x}$

  Identity 6 $\sin^2 x + \cos^2 x = 1$
  Identity 7 $\tan^2 x + 1 = \sec^2 x$
  Identity 8 $\cot^2 x + 1 = \csc^2 x$
  Identities 6, 7, and 8 are called Pythagorean identities.

- There is no standard procedure for working with identities. In fact, a given identity can usually be proved in several ways. However, the following suggestions should be helpful.

  1. Change the more complicated expression in the identity to the same form as the less complicated expression.
  2. Until you become more familiar with identities, change all functions to sines and cosines. This procedure might necessitate more algebra in some instances, but it will provide a direct approach to the problem. Gradually try to make use of the other trigonometric functions.
  3. Do not attempt to prove an identity by treating it as an equation and using the associated techniques, for this involves assuming what you want to prove.

Detailed solutions to selected exercises

Exercise 18   Show $\sin x(\csc x - \sin x)$ is identical to $\cos^2 x$.

Solution $\sin x(\csc x - \sin x) = \sin x \csc x - \sin^2 x$
$$= 1 - \sin^2 x \qquad \text{Identity 1}$$
$$= \cos^2 x \qquad \text{Identity 6}$$

Now do **A**.

Exercise 24   Prove the identity $(1 + \tan^2 x)/\tan^2 x = \csc^2 x$.

Solution $\dfrac{1 + \tan^2 x}{\tan^2 x} = \dfrac{1 + (\sin^2 x/\cos^2 x)}{\sin^2 x/\cos^2 x}$ Identity 4

$$= \dfrac{\cos^2 x + \sin^2 x}{\sin^2 x} \qquad \text{Simplify the complex fraction.}$$

$$= \dfrac{1}{\sin^2 x} \qquad \text{Identity 6}$$

$$= \csc^2 x \qquad \text{Identity 1}$$

Alternative solution   Using $\tan^2 x + 1 = \sec^2 x$ (Identity 7), the given expression becomes

$$\frac{\sec^2 x}{\tan^2 x} = \frac{1/\cos^2 x}{\sin^2 x/\cos^2 x} = \frac{1}{\sin^2 x} = \csc^2 x.$$

Now do **B**.

Exercise 38   Prove the identity $\sec^4 x - \tan^4 x = 2 \sec^2 x - 1$.

Solution $\sec^4 x - \tan^4 x = (\sec^2 x + \tan^2 x)(\sec^2 x - \tan^2 x)$
$$= [\sec^2 x + (\sec^2 x - 1)][\sec^2 x - (\sec^2 x - 1)]$$
$$= 2 \sec^2 x - 1$$

Now do **C**.

B. Prove the identity
$(1 + \cot^2 x)/\cot^2 x = \sec^2 x.$

C. Prove the identity
$\tan^4 x - \sec^4 x$
$= -(\tan^2 x + \sec^2 x).$

6.8 More on trigonometric identities

Objectives checklist

Can you:

a. State the identities for the sine, cosine, and tangent of the sum or difference of two numbers and use these formulas to verify certain other important identities?

b. State the double angle identities for $\sin 2x$ and $\cos 2x$ and verify identities involving these expressions?

c. Use identities to evaluate various trigonometric expressions given $\sin x$ and/or $\cos x$ and the interval containing x?

Key rules and formulas

- The key identities in this section are as follows:
  Sum or difference formulas

  1. $\cos(x_1 + x_2) = \cos x_1 \cos x_2 - \sin x_1 \sin x_2$
  2. $\cos(x_1 - x_2) = \cos x_1 \cos x_2 + \sin x_1 \sin x_2$
  3. $\sin(x_1 + x_2) = \sin x_1 \cos x_2 + \cos x_1 \sin x_2$
  4. $\sin(x_1 - x_2) = \sin x_1 \cos x_2 - \cos x_1 \sin x_2$
  5. $\tan(x_1 + x_2) = \dfrac{\tan x_1 + \tan x_2}{1 - \tan x_1 \tan x_2}$

  6. $\tan(x_1 - x_2) = \dfrac{\tan x_1 - \tan x_2}{1 + \tan x_1 \tan x_2}$

  Double-angle formulas
  7. $\sin 2x = 2 \sin x \cos x$
  8. a. $\cos 2x = \cos^2 x - \sin^2 x$
     b. $\cos 2x = 2 \cos^2 x - 1$
     c. $\cos 2x = 1 - 2 \sin^2 x$
  9. $\tan 2x = \dfrac{2 \tan x}{1 - \tan^2 x}$

Additional comments

- For reference purposes, a summary of the major identities appears at the end of Section 6.8 in the text.

6.8

A. Prove the identity
$\cos\left(\frac{1}{2}x - \pi/2\right) = \sin\frac{1}{2}x.$

B. If $\cos x_1 = -\frac{12}{13}$, where
$\pi/2 < x_1 < \pi$, and $\sin x_2 = -\frac{3}{5}$
where $\pi < x_2 < 3\pi/2$, find
 a. $\sin x_1$
 b. $\cos x_2$
 c. $\cos(x_1 - x_2)$
 d. $\sin(x_1 + x_2)$

C. Use the formula for the sine of
the sum of two numbers to verify
that $\sin 3x = 3\sin x - 4\sin^3 x$
is an identity.

Detailed solutions to selected exercises

Exercise 4 Prove the identity $\cos\left(\dfrac{3\pi}{2} - x\right) = -\sin x.$

Solution Using the formula for the cosine of the difference of two numbers (identity 2), we have

$$\cos\left(\frac{3\pi}{2} - x\right) = \cos\frac{3\pi}{2}\cos x + \sin\frac{3\pi}{2}\sin x$$
$$= (0)\cos x + (-1)\sin x$$
$$= -\sin x.$$

Now do **A**.

Exercise 16 See text for question.

Solution We start with $\cos 2x = 1 - 2\sin^2 x$ and (following the method outlined in text Example 5) we replace x by $x/2$ to obtain $\cos 2(x/2) = 1 - 2\sin^2(x/2)$. Then solving for $\sin^2(x/2)$, we have
$\sin^2(x/2) = \dfrac{1 - \cos x}{2}$, so $\sin(x/2) = \pm\sqrt{\dfrac{1 - \cos x}{2}}$.

Exercise 24 See text for question.

Solution

 a. Since $3\pi/2 < x_1 < 2\pi$, $\sin x_1$ is negative, so
$$\sin x_1 = -\sqrt{1 - \cos^2 x_1} = -\sqrt{1 - (3/5)^2} = -\sqrt{16/25} = -\tfrac{4}{5}.$$

 b. Since $3\pi/2 < x_2 < 2\pi$, $\cos x_2$ is positive, so
$$\cos x_2 = \sqrt{1 - \sin^2 x_2} = \sqrt{1 - (-5/13)^2} = \sqrt{144/169} = \tfrac{12}{13}.$$

 c. $\cos(x_1 - x_2) = \cos x_1 \cos x_2 + \sin x_1 \sin x_2$
$$= (\tfrac{3}{5})(\tfrac{12}{13}) + (\tfrac{-4}{5})(\tfrac{-5}{13})$$
$$= \tfrac{56}{65}$$

 d. $\sin(x_1 + x_2) = \sin x_1 \cos x_2 + \cos x_1 \sin x_2$
$$= (\tfrac{-4}{5})(\tfrac{12}{13}) + (\tfrac{3}{5})(\tfrac{-5}{13})$$
$$= -\tfrac{63}{65}$$

Now do **B**.

Exercise 26a Use the formula for the cosine of the sum of two numbers to verify the identity $\cos 3x = 4\cos^3 x - 3\cos x$.

Solution We write $\cos 3x$ as $\cos(2x + x)$ and use the formula for the cosine of the sum of two numbers.

$$\cos 3x = \cos(2x + x) = \cos 2x \cdot \cos x - \sin 2x \cdot \sin x$$

Next if we apply the double-angle formulas $\cos 2x = 2\cos^2 x - 1$ and $\sin 2x = 2\sin x \cos x$, we have

$$\cos 3x = (2\cos^2 x - 1)\cos x - (2\sin x \cos x)\sin x$$
$$= 2\cos^3 x - \cos x - 2\cos x \sin^2 x.$$

Finally, we apply the Pythagorean identity $1 - \cos^2 x = \sin^2 x$ and simplify.

$$\cos 3x = 2\cos^3 x - \cos x - 2\cos x(1 - \cos^2 x)$$
$$= 2\cos^3 x - \cos x - 2\cos x + 2\cos^3 x$$
$$= 4\cos^3 x - 3\cos x.$$

Now do **C**.

Exercise 34   Prove the identity $\cos 2x/\sin x + \sin 2x/\cos x = \csc x$.

Solution   First, use the double-angle identities for $\sin 2x$ and $\cos 2x$.

$$\frac{\cos 2x}{\sin x} + \frac{\sin 2x}{\cos x} = \frac{1 - 2\sin^2 x}{\sin x} + \frac{2\sin x \cos x}{\cos x}$$

Dividing by the common factor, $\cos x$, and adding gives

$$\frac{1 - 2\sin^2 x + 2\sin^2 x}{\sin x} = \frac{1}{\sin x} = \csc x.$$

Now do **D.**

D. Prove the identity
$$\frac{1 - \cos 2x}{1 + \cos 2x} = \tan^2 x.$$

6.9 Trigonometric equations

Objectives checklist

Can you:

a. Find the exact solution to trigonometric equations in which the answers are quadrantal angles or angles with reference angles of $\pi/6$, $\pi/4$, or $\pi/3$?

b. Find the approximate solution to trigonometric equations using a calculator or a table and reference angles?

c. Use factoring or the quadratic formula to solve certain trigonometric equations?

d. Use identities to help solve certain trigonometric equations?

Key terms

Trigonometric identity   Conditional trigonometric equation

Key rules and formulas

•   In solving a trigonometric equation, we are looking for the set of all values of the variable (in a certain interval) that satisfy the equation. Solutions that are quadrantal angles may be read from the chart on the back endpaper of the text. Otherwise, follow these steps.

1. If necessary, solve the equation for the functions of x in the problem. Identities, factoring, and the quadratic formula may be needed here.
2. Determine two quadrants for the point assigned to x by using the chart on the "Signs of the Trigonometric Ratios." (See back endpaper in text or chart from Section 6.4.)
3. Find the reference angle or number.
4. Determine solutions between 0 and 2π as follows:

Quadrant	Solution
1	reference angle
2	π − reference angle
3	π + reference angle
4	2π − reference angle

5. If necessary, write the formula that generates angles coterminal to the above solutions. In general, if x is a solution, then $x + k2\pi$, where k is any integer, is also a solution of the equation.

6.9

A. Solve $\sqrt{3} \sec x + 2 = 0$
$(0 \leq x < 2\pi)$.

B. Solve $2 \cot x - 6 = 0$ (all solutions).

C. Solve on the interval $[0,2\pi)$:
$2 \sin^2 2x + 3 \sin 2x - 2 = 0$.

Additional Comments

- Be sure to set the calculator for radian mode when solving these problems. Most calculators are in degree mode when first turned on, and forgetting this step is a very common error. For other notes on calculator usage, consult the text.

Detailed solutions to selected exercises

Exercise 12 Solve $\csc x - \sqrt{2} = 0$ $(0 \leq x < 2\pi)$.

Solution We follow the steps outlined:

1. $\csc x - \sqrt{2} = 0$ so $\csc x = \sqrt{2}$.
2. $\csc x$ is positive in quadrants 1 and 2.
3. The reference angle is $\pi/4$ (from chart on endpaper of text or $1{:}1{:}\sqrt{2}$ ratios in 45-45-90 triangle).
4. Q_1 solution: reference angle $= \pi/4$
   Q_2 solution: $\pi -$ reference angle $= \pi - \pi/4 = 3\pi/4$

Thus, $\pi/4$ and $3\pi/4$ make the equation a true statement, so the solution set is $\left\{ \dfrac{\pi}{4}, \dfrac{3\pi}{4} \right\}$.

Now do **A.**

Exercise 20 Solve $5 \cot x + 3 = 0$ (all solutions).

Solution We follow the steps outlined:

1. $5 \cot x + 3 = 0$ so $\cot x = -0.6$
2. $\cot x$ is negative in quadrants 2 and 4.
3. The reference number is 1.03 (from Table 6 in the text or 0.6 $\boxed{1/x}$ $\boxed{INV}$ $\boxed{\tan}$ on a calculator).
4. Q_2 solution: $\pi - 1.03 = 3.14 - 1.03 = 2.11$
   Q_4 solution: $2\pi - 1.03 = 6.28 - 1.03 = 5.25$
5. The solution set is $\{x\colon x = 2.11 + k2\pi$ or $x = 5.25 + k2\pi, k$ any integer$\}$.

Now do **B.**

Exercise 36 Solve $\cot^2 2x + 2 \cot 2x + 1 = 0$ $(0 \leq x < 2\pi)$.

Solution First, factor the left side of the equation.

$$\cot^2 2x + 2 \cot 2x + 1 = (\cot 2x + 1)(\cot 2x + 1) = 0$$

Now set $\cot 2x + 1 = 0$ and solve for $2x$ in the usual way.

1. $\cot 2x + 1 = 0$ so $\cot 2x = -1$.
2. $\cot x$ is negative in quadrants 2 and 4.
3. The reference angle is $\pi/4$ (from chart on endpaper in text or $1{:}1{:}\sqrt{2}$ ratios in 45-45-90 triangle).
4. Q_2 solution: $\pi - \pi/4 = 3\pi/4$
   Q_4 solution: $2\pi - \pi/4 = 7\pi/4$
5. All solutions: $2x = 3\pi/4 + k2\pi$ so $x = 3\pi/8 + k\pi$
   $2x = 7\pi/4 + k2\pi$ so $x = 7\pi/8 + k\pi$

Finally, to obtain solutions in the interval $[0,2\pi)$, use 0 and 1 as replacements for k. Thus, the solution set is $\{3\pi/8, 7\pi/8, 11\pi/8, 15\pi/8\}$.
Now do **C.**

Exercise 38   Use the quadratic formula to solve $3 \sin^2 x - \sin x - 1 = 0$ in the interval $0 \le x < 2\pi$.

Solution   In this equation $a = 3$, $b = -1$, and $c = -1$. Therefore,

$$\sin x = \frac{-(-1) \pm \sqrt{(-1)^2 - 4(3)(-1)}}{2(3)} = \frac{1 \pm \sqrt{13}}{6} \quad \text{so}$$

$$\sin x = 0.7676 \text{ or } -0.4343.$$

For $\sin x = 0.7676$, Q_1 solution: 0.88, Q_2 solution: $\pi - 0.88 = 2.26$.
For $\sin x = -0.4343$, Q_3 solution: $\pi + 0.45 = 3.59$, Q_4 solution:
$2\pi - 0.45 = 5.83$.
Thus, $\{0.88, 2.26, 3.59, 5.83\}$ is the solution set of the equation for $0 \le x < 2\pi$.
Now do **D**.

Exercise 50   Use identities and solve $1 - \cos^2 x = \sin x$ for $0 \le x < 2\pi$.

Solution $1 - \cos^2 x = \sin x$
$\qquad\qquad \sin^2 x = \sin x \qquad$ Using $\sin^2 x + \cos^2 x = 1$
$\qquad \sin^2 x - \sin x = 0$
$\quad \sin x(\sin x - 1) = 0$

Now set each factor to zero and use the chart on the endpaper of the text.

$$\sin x = 0 \text{ when } x = 0 \text{ and } \pi, \text{ while } \sin x = 1 \text{ when } x = \pi/2$$

Thus, $\{0, \pi/2, \pi\}$ is the solution set of the equation for $0 \le x < 2\pi$.
Now do **E**.

D. Use the quadratic formula to solve $5 \sin^2 x - \sin x - 2 = 0$ in the interval $0 \le x < 2\pi$.

E. Use identities and solve $\cos^2 x - \sin^2 x = 1$ for $0 \le x < 2\pi$.

6.10 Inverse trigonometric functions

Objectives checklist

Can you:

a. State the domain and range of the six inverse trigonometric functions?

b. Evaluate expressions involving the inverse trigonometric functions by using a calculator, a table, or a right triangle?

c. Express certain relationships as a function of an indicated variable by using inverse trigonometric functions and/or right triangles?

d. Solve a given formula for θ or t by using inverse trigonometric functions?

Key rules and formulas

- Keep in mind the following about any inverse functions:

  1. Two functions with exactly reverse assignments are inverse functions.
  2. The domain of a function f is the range of its inverse function and the range of f is the domain of its inverse.
  3. A function is one-to-one when each x value in the domain is assigned a different y value so that no two ordered pairs have the same second component. We define the inverse of f, denoted f^{-1}, only when f is a one-to-one function.

- The inverse of a trigonometric function, say $y = \sin x$, is written as $y = \arcsin x$ or $y = \sin^{-1} x$. Both expressions are read "y is the angle (or number) whose sine is x." By definition,

$$y = \arcsin x \text{ if and only if } x = \sin y.$$

6.10

A. Evaluate $\arcsin(-\frac{1}{2})$.

B. Evaluate $\cos[\sin^{-1}(-\sqrt{2}/2)]$.

The domain and range for each inverse trigonometric function is as follows:

Function	Domain	Range
$y = \arcsin x$	$\{x: -1 \leq x \leq 1\}$	$\left\{y: -\dfrac{\pi}{2} \leq y \leq \dfrac{\pi}{2}\right\}$
$y = \arccos x$	$\{x: -1 \leq x \leq 1\}$	$\{y: 0 \leq y \leq \pi\}$
$y = \arctan x$	Set of all real numbers	$\left\{y: -\dfrac{\pi}{2} < y < \dfrac{\pi}{2}\right\}$
$y = \text{arccot } x$	Set of all real numbers	$\{y: 0 < y < \pi\}$
$y = \text{arcsec } x$	$\{x: x \leq -1 \text{ or } x \geq 1\}$	$\left\{y: 0 \leq y \leq \pi, y \neq \dfrac{\pi}{2}\right\}$
$y = \text{arccsc } x$	$\{x: x \leq -1 \text{ or } x \geq 1\}$	$\left\{y: -\dfrac{\pi}{2} \leq y \leq \dfrac{\pi}{2}, y \neq 0\right\}$

Additional comments

- In terms of notation, an inverse trigonometric function is sometimes capitalized and written Arcsin x or Sin^{-1} x. With this convention, arcsin x and sin^{-1} x do not represent functions but represent the set of all numbers whose sine is x with no restriction placed on the range values.

- The inverse secant and cosecant functions are sometimes assigned different range intervals depending on the application for the function.

Detailed solutions to selected exercises

Exercise 4   Evaluate arcsin $(\sqrt{3}/2)$.

Solution   arcsin $(\sqrt{3}/2)$ is the number between $-\pi/2$ and $\pi/2$, inclusive, whose sine is $\sqrt{3}/2$. From the chart on the endpaper or from the $1:\sqrt{3}:2$ ratios in a 30-60-90 triangle, we determine the number is $\pi/3$.
Now do **A.**

Exercise 24   Evaluate $\cos[\sin^{-1}(-\frac{1}{2})]$.

Solution   First, in $[-\pi/2, \pi/2]$ the number whose sine is $-\frac{1}{2}$ is $-\pi/6$. Then $\cos[\sin^{-1}(-\frac{1}{2})] = \cos(-\pi/6) = \cos \pi/6 = \sqrt{3}/2$.
By calculator: 0.5 $\boxed{+/-}$ $\boxed{\text{INV}}$ $\boxed{\text{sin}}$ $\boxed{\text{cos}}$ 0.8660
Now do **B.**

Exercise 34   Evaluate $\cot (\cos^{-1} \frac{8}{17})$ using right triangles.

Solution $\cos^{-1} \frac{8}{17}$ is the angle whose cosine is $\frac{8}{17}$. Label this angle θ and draw a right triangle as in Figure 6.10:34 with hypotenuse of length 17 and adjacent side of length 8. By the Pythagorean theorem, the length of the opposite side is then 15. Finally,

$$\cot(\cos^{-1} \tfrac{8}{17}) = \cot \theta = \frac{\text{adjacent}}{\text{opposite}} = \frac{8}{15}.$$

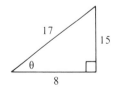

Figure 6.10:34

Now do **C**.

Exercise 46 If $x - 1 = \sin \theta$, write $\theta + \sin \theta \cos \theta$ as a function of x.

Solution If $x - 1 = \sin \theta$, then $\theta = \arcsin(x - 1)$. We can represent $\cos \theta$ by interpreting θ as the measure of a right triangle as shown in Figure 6.10:46. The length of the opposite side is $x - 1$, the length of the hypotenuse is 1, and the length of the adjacent side is found by the Pythagorean theorem to be $\sqrt{1 - (x - 1)^2}$. Thus, $\cos \theta = \sqrt{1 - (x - 1)^2}/1$ so

$$\theta + \sin \theta \cos \theta = \arcsin(x - 1) + (x - 1)\sqrt{1 - (x - 1)^2}.$$

D. If $x + 1 = \cos \theta$, write $\theta - \sin \theta \cos \theta$ as a function of x.

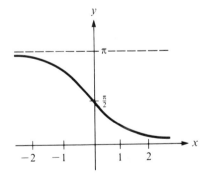

Figure 6.10:46

Now do **D**.

E. Solve $a = \cos \pi t$ for t.

Exercise 52 Solve $a = \sin \pi t$ for t.

Solution $a = \sin \pi t$, so $\pi t = \arcsin a$, and $t = (\arcsin a)/\pi$.
Now do **E**.

Exercise 56 Sketch the graph of $y = \text{arccot } x$.

Solution The domain of $y = \text{arccot } x$ is the set of all real numbers, while the range is $\{y: 0 < y < \pi\}$. Based on this information and the table below, we graph this function as in Figure 6.10:56.

x	-2	-1	0	1	2
y	2.6779	$\dfrac{3\pi}{4}$	$\dfrac{\pi}{2}$	$\dfrac{\pi}{4}$	0.4636

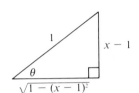

Figure 6.10:56

Caution! One troublesome spot on a calculator is evaluating the inverse cotangent of a negative number. For this function we define $\pi/2 < y < \pi$ for $x < 0$. But the inverse tangent function is defined and is programmed for $-\pi/2 < y < 0$ when $x < 0$. Since the two functions do not have the same range, you should use the calculator to find the reference number and then determine the solution between $\pi/2 < y < \pi$ as discussed in Section 6.9. For example, to determine $\text{arccot}(-2)$, we first determine that the reference number is 0.4636. Then $\text{arccot}(-2) = \pi - (0.4636) = 2.6779$.

Exercise 58   For what values of x is the statement $\sin(\sin^{-1} x) = x$ true?

Solution   By definition, $\sin^{-1} x = y$ if and only if $\sin y = x$, where $-1 \le x \le 1$ and $-\pi/2 \le y \le \pi/2$. Thus, $\sin(\sin^{-1} x) = \sin y = x$ for all values of x in the interval $[-1,1]$.

Sample test questions: Chapter 6

1. Evaluate each expression. Give exact answers where possible.

   a. $\sin 60°$ b. $\sec 95°$ c. $\tan 302°10'$

   d. $\cos(-45°)$ e. $\cot 90°$ f. $\csc 400°$

   g. $\sin 3\pi$ h. $\sin 3$ i. $\tan 1$

   j. $\cos(-\pi/3)$ k. $\sin^{-1} 0.3124$ l. $\csc 0$

2. Complete the statement.

   a. If $(3,-1)$ is on the terminal ray of θ, then $\tan \theta$ equals _____ .

   b. If $\sin \theta = \frac{2}{3}$, then $\csc \theta$ equals _____ .

   c. If $\tan \theta = 1/\sqrt{3}$ and $\sin \theta < 0$, then $\cos \theta$ equals _____ .

   d. The cosine function is negative in quadrants _____ .

   e. If $\tan A = \frac{7}{12}$ and $a = 21$ in right triangle ABC, then b equals _____ .

   f. An angle coterminal to $10°$ is _____ .

   g. If $\sin 23° = \cos \theta$ and θ is an acute angle, then θ equals _____ .

   h. The reference angle for $280°$ is _____ .

   i. The domain of $y = \tan x$ is _____ .

   j. The range of $y = \arcsin x$ is _____ .

   k. The period of $y = 2 \cos 3x$ is _____ .

   l. The reference angle for $7\pi/5$ is _____ .

   m. $330°$ expressed in radians is _____ .

   n. $3\pi/4$ expressed in degrees is _____ .

   o. The function $y = 2 \sin 4x$ attains a maximum value for an acute angle of _____ .

   p. The area of a sector of a circle whose central angle is $120°$ and whose radius is 6 in. is _____ .

   q. If $a = \sin(x - 1)$, then x equals _____ .

3. Solve each equation for the specified interval to the nearest degree.

   a. $\tan \theta = 1.234$ $(0° \le \theta \le 90°)$

   b. $\cos \theta = 1/\sqrt{2}$ $(270° \le \theta \le 360°)$

   c. $2 \cos \theta - 2 = 0$ (all solutions)

   d. $\tan 2\theta = \cot(2\theta - 14°)$ $(0° < \theta < 90°)$

   e. $\dfrac{\sin 30°}{6.0} = \dfrac{\sin \theta}{5.0}$ $(0° < \theta < 180°)$

   f. $361 = 225 + 121 - 2(11)(15) \cos \theta$ (for $0° < \theta < 180°$)

4. Solve each equation for the specified interval of real numbers.

    a. $2 \tan x + 7 = 0$ (all solutions)

    b. $\cos 2x = -\frac{1}{2}$ $(0 \leq x < \pi)$

    c. $\sin^2 x = \sin x$ $(0 \leq x < 2\pi)$

    d. $\tan x + \cot x = \sqrt{2} \csc x$ $(0 \leq x < 2\pi)$

5. The distance from ground level to the underside of a cloud is called the "ceiling." At an airport, a ceiling light projector throws a spotlight vertically on the underside of a cloud. At a distance of $60\overline{0}$ ft from the projector, the angle of elevation of the spot of light on the cloud is $60.0°$. What is the ceiling?

6. Verify each identity.

    a. $\cot^2 x + 1 = \csc^2 x$ b. $\sin^2 x = \dfrac{1 - \cos 2x}{2}$

    c. $\cos\left(\dfrac{3\pi}{2} + x\right) = \sin x$

7. If $x = \sin \theta$, write $\theta + \sin 2\theta$ as a function of x.

8. If $\sin x = a$ and $0 < x < \pi/2$, find

    a. $\csc x$ b. $\cos 2x$ c. $\arcsin a$

9. Graph each function.

    a. $y = \frac{1}{2} \sin 3x;\ 0 \leq x \leq 2\pi$

    b. $y = -\cos\left(\dfrac{x}{4} + \dfrac{\pi}{2}\right)$; one cycle

    c. $y = \arcsin x$

10. Select the choice that completes the statement or answers the question.

    a. Which number is not a possible value for $\sin \theta$?

        (a) 2   (b) $\frac{1}{2}$ (c) 0   (d) 1   (e) -0.9

    b. $\cos 200°$ simplifies to

        (a) $\cos 20°$ (b) $-\cos 20°$ (c) $\cos 70°$ (d) $-\cos 70°$

    c. If $\sec \theta = k$, then $\cos \theta$ equals

        (a) k (b) $-k$ (c) $-1/k$ (d) $1/k$ (e) $1 - k$

    d. If the hypotenuse in a right triangle is 15 and one leg is 5.0, then the *smallest* angle in the triangle is

        (a) $19°30'$ (b) $90°$ (c) $70°30'$ (d) $18°20'$ (e) $21°40'$

    e. Which number is not in the range of $y = \csc x$?

        (a) 1   (b) 0   (c) 2   (d) -2 (e) $\sqrt{2}$

    f. Which one of the following is an identity?

        (a) $\sin x + \cos x = 1$ (b) $\sin x \cdot \cos x = 1$

        (c) $\sin 2x = 2 \sin x$ (d) $\sin(-x) = -\sin x$

        (e) $\cos 2x = 2 \sin x \cos x$

g. A function having the period 2 is

(a) $y = 2 \cos x$ (b) $y = \cos 2x$ (c) $y = \cos(x/2)$

(d) $y = \dfrac{\cos x}{2}$ (e) $y = \cos \pi x$

h. The phase shift for $y = 2 \sin(\pi x - \pi)$ is

(a) 1   (b) -1 (c) 2   (d) π (e) $-\pi$

i. The expression $\sin x \cdot \sec x \cdot \cot x$ is identical to

(a) 0   (b) 1   (c) $\tan x$ (d) $\cos x$ (e) $\csc x$

Chapter 7
Trigonometry: General Angle Applications

7.1 Law of sines

Objectives checklist

Can you:

a. Use the law of sines to solve any triangle given two angles and one side of the triangle?

b. Use the law of sines to solve any triangle given two sides of the triangle and the angle opposite one of them?

c. Solve applied problems using the law of sines?

Key rules and formulas

- *Law of sines:* The sines of the angles in a triangle are proportional to the lengths of the opposite sides. In symbols,

$$\frac{\sin A}{a} = \frac{\sin B}{b} = \frac{\sin C}{c} \quad \text{or} \quad \frac{a}{\sin A} = \frac{b}{\sin B} = \frac{c}{\sin C}.$$

Additional comments

- Guidelines for the desired accuracy in a solution are the same as in Section 6.1.

- When given the measures for two sides of a triangle and the angle opposite one of them, there may be one triangle that fits the data (see text, Example 3) or there may be two triangles that fit the data (see text, Example 4). If no triangle can be constructed from the data, then the data is inconsistent.

- Computed results should not be used to determine the measures for other parts in a triangle when it is possible to use given data.

Detailed solutions to selected exercises

Exercise 4   Solve the triangle in which $C = 135°$, $c = 98$ ft, and $B = 15°$.

Solution   First, sketch Figure 7.1:4. Since $A + B + C = 180°$, we have

$$A = 180° - (15° + 135°) = 30°.$$

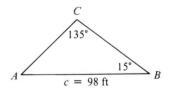

Figure 7.1:4

Chapter 7

7.1

A. Solve the triangle in which $C = 120°$, $c = 49$ ft, and $B = 7°$.

By the law of sines, $a/\sin A = c/\sin C$. Thus,

$$\frac{a}{\sin 30°} = \frac{98}{\sin 135°} \quad \text{so} \quad a = \frac{98 \sin 30°}{\sin 135°} = 69.$$

In a similar way, $b/\sin B = c/\sin C$. Then

$$\frac{b}{\sin 15°} = \frac{98}{\sin 135°} \quad \text{so} \quad b = \frac{98 \sin 15°}{\sin 135°} = 36.$$

Therefore, $A = 30°$, $a = 69$ ft, and $b = 36$ ft.
Now do **A.**

Exercise 12   Solve the triangle in which $C = 105°30'$, $c = 46.1$ ft, and $b = 75.2$ ft.

B. Solve the triangle in which $C = 95°30'$, $c = 28.3$ ft, and $b = 67.5$ ft.

Solution   No triangle satisfies the data. We can reason that since $b > c$, $B > C$, which is impossible since a triangle cannot have more than one obtuse angle. In terms of the law of sines, $(\sin B)/b = (\sin C)/c$ so

$$\frac{\sin B}{75.2} = \frac{\sin 105°30'}{46.1} \quad \text{yielding}$$

$$\sin B = \frac{75.2 \sin 105°30'}{46.1} = 1.572.$$

Since $\sin B$ cannot exceed 1, the data is inconsistent.
Now do **B.**

C. The short side of a parallelogram measures 75.0 ft and the shorter diagonal measures 98.0 ft. Find the length of the longer side if the angle between the longer side and the shorter diagonal measures 39.0°.

Exercise 16   See text for question.

Solution   First, sketch Figure 7.1:16. To find A, we use $(\sin A)/a = (\sin B)/b$ as follows:

$$\frac{\sin A}{700} = \frac{\sin 25°}{800} \quad \text{so} \quad \sin A = \frac{700 \sin 25°}{800} = 0.3698 \text{ yielding } A = 21.7°.$$

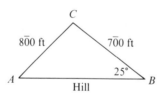

Figure 7.1:16

We reject the obtuse case for angle A, since $a < b$. Now since $A + B + C = 180°$, we know

$$C = 180° - (25° + 21.7°) \quad \text{so} \quad C = 133.3°.$$

Finally, by the law of sines, $c/\sin C = b/\sin B$. Then

$$\frac{c}{\sin 133.3°} = \frac{800}{\sin 25°} \quad \text{so} \quad c = \frac{800 \sin 133.3°}{\sin 25°} = 1,378.$$

To two significant digits, the distance is 1,400 ft.
Now do **C.**

7.2 Law of cosines

Objectives checklist

Can you:

a. Use the law of cosines to solve any triangle given the lengths of three sides of the triangle?

b. Use the law of cosines, and then the law of sines, to solve any triangle given the measures for two sides of the triangle and the angle between these two sides?

c. Determine whether the law of sines or the law of cosines is appropriate for solving a given problem?

d. Solve applied problems using the law of cosines?

Key rules and formulas

- *Law of cosines:* In any triangle the square of any side length equals the sum of the squares of the other two side lengths, minus twice the product of these other two side lengths and the cosine of their included angle. In symbols,

$$a^2 = b^2 + c^2 - 2bc \cos A$$
$$b^2 = a^2 + c^2 - 2ac \cos B$$
$$c^2 = a^2 + b^2 - 2ab \cos C.$$

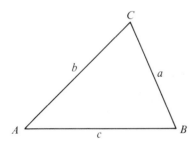

When the lengths of three sides of the triangle are known, it is convenient to restate these formulas as

$$\cos A = \frac{b^2 + c^2 - a^2}{2bc} \; ; \cos B = \frac{a^2 + c^2 - b^2}{2ac} \; ; \cos C = \frac{a^2 + b^2 - c^2}{2ab} \; .$$

Detailed solutions to selected exercises

Exercise 10 Solve the triangle in which $a = 45.0$ ft, $b = 108$ ft, and $c = 117$ ft.

Solution By the law of cosines,

$$\cos A = \frac{b^2 + c^2 - a^2}{2bc} = \frac{(108)^2 + (117)^2 - (45.0)^2}{2(108)(117)} = 0.9231.$$

$A = 22°40'$ or $22.6°$ (from Table 5 or by calculator: 0.9231 $\boxed{\text{INV}}$ $\boxed{\text{cos}}$). We find B in a similar way.

$$\cos B = \frac{a^2 + c^2 - b^2}{2ac} = \frac{(45.0)^2 + (117)^2 - (108)^2}{2(45.0)(117)} = 0.3846$$

$B = 67°20'$ or $67.4°$. Finally, since $A + B + C = 180°$,

$$C = 180° - (22.6° + 67.4°) = 90°00' \text{ or } 90.0°.$$

7.2

A. Solve the triangle in which $a = 30.0$ ft, $b = 75.0$ ft, and $c = 96.0$ ft.

B. Solve the triangle in which $b = 99.9$ ft, $c = 70.3$ ft, and $A = 60°30'$.

C. One gun is located at point A, while a second gun at point B is located 25 miles directly east of A. From point A the direction to the target is 43° north of east. From point B the direction to the target is 72° north of west. For what firing range should the guns be set?

In fact, ABC is a right triangle, since $117^2 = 45^2 + 108^2$.
Now do **A.**

Exercise 18 Solve the triangle in which $b = 126$ ft, $c = 92.1$ ft, and $A = 72°50'$.

Solution First, sketch Figure 7.2:18. We know the angle measure between two given sides so we use the law of cosines to find side length a.

$$a^2 = b^2 + c^2 - 2bc \cos A$$
$$= (126)^2 + (92.1)^2 - 2(126)(92.1) \cos 72°50' = 17,508$$
$$a = \sqrt{17,508} = 132.3$$

(*Note:* To three significant digits, $a = 132$ ft.)
Now use the law of sines to find the smaller unknown angle (which must be acute). From $(\sin C)/c = (\sin A)/a$, we have

$$\frac{\sin C}{92.1} = \frac{\sin 72°50'}{132.3} \quad \text{so} \quad \sin C = \frac{92.1 \sin 72°50'}{132.3} = 0.6651$$
$$\text{so } C = 41°40' \text{ or } 41.7°.$$

Finally, since $A + B + C = 180°$,

$$B = 180° - (72°50' + 41°40') = 65°30'.$$

Now do **B.**

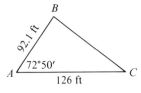

Figure 7.2:18

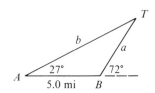

Figure 7.2:28

Exercise 28 See text for question.

Solution First, sketch Figure 7.2:28. Since $B + 72° = 180°$, $B = 108°$. Since $A + B + T = 180°$, $T = 45°$. All the angle measures in the triangle are known so we find side lengths a and b by applying the law of sines.

$$\frac{a}{\sin 27°} = \frac{5.0}{\sin 45°} \quad \text{so} \quad a = \frac{5 \sin 27°}{\sin 45°} = 3.2 \text{ mi}$$
$$\frac{b}{\sin 108°} = \frac{5.0}{\sin 45°} \quad \text{so} \quad b = \frac{5 \sin 108°}{\sin 45°} = 6.7 \text{ mi}$$

Now do **C.**

7.3 Vectors

Objectives checklist

Can you:

a. Find the resultant, or vector sum, of two vectors by using triangle trigonometry?

b. Resolve vectors into components?

c. Find the resultant, or vector sum, of given vectors by using components?

d. Solve applied problems using vectors?

Key terms

Vector
Resultant
Vector sum

Components of a vector
Resolving a vector

Key rules and formulas

• Quantities that have both magnitude and direction are called vectors. When two forces, vectors **OA** and **OB,** act on a body, the resultant is vector **OC,** which is the diagonal of a parallelogram formed from the given vectors.

• The resultant of two (or more) vectors may be found by breaking each vector into its x and y components. The sum of the x components of each vector is the x component of the resultant (labeled R_x). Similarly, the sum of the y components of each vector is the y component of the resultant (labeled R_y). The magnitude of the resultant and the angle the resultant makes with the positive x-axis (labeled θ_R) are determined by the formulas

$$R = \sqrt{(R_x)^2 + (R_y)^2} \quad \text{and} \quad \tan \theta_R = R_y/R_x.$$

Detailed solutions to selected exercises

Exercise 8   See text for question.

Solution   First, sketch Figure 7.3:8. The magnitude of vector **OB** is $32(5) = 160$ ft/second and $\overline{OB} = \overline{AC}$. By the Pythagorean theorem, $(\overline{OC})^2 = (\overline{OA})^2 + (\overline{AC})^2$. Thus,

$$(\overline{OC})^2 = (300)^2 + (160)^2 = 115{,}600 \text{ so } \overline{OC} = \sqrt{115{,}600} = 340 \text{ ft/sec.}$$

We find angle θ by using the tangent function.

$$\tan \theta = \frac{160}{300} = 0.5333 \quad \text{so} \quad \theta = 28°$$

Now do **A.**

7.3

A. An object is dropped from a plane that is moving horizontally at a speed of $24\overline{0}$ ft/sec. If the vertical velocity of the object in terms of time is given by the formula $v = 32t$, 10 seconds later:
   a. what is the speed of the object?
   b. what angle does the direction of the object make with the horizontal?

B. Find the horizontal and vertical components of a vector that has a magnitude of 5.0 lb and makes an angle of 40° with the positive x-axis.

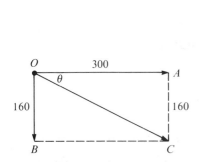

Figure 7.3:8

Figure 7.3:14

Exercise 14   See text for question.

Solution   First, sketch Figure 7.3:14. Then

Horizontal component: $\cos 72° = x/125$ so $x = 125 \cos 72° = 38.6$ lb
Vertical component: $\sin 72° = y/125$ so $y = 125 \sin 72° = 119$ lb.

Now do **B.**

C. Vector **A** has a magnitude of $10\overline{0}$ lb and makes an angle of 20° with the positive x-axis. Vector **B** has a magnitude of 85 lb and makes an angle of 40° with the positive x-axis. Vector **C** has a magnitude of 9.0 lb and makes an angle of 160° with the positive x-axis. Find the resultant (or vector sum) of **A**, **B**, and **C** by resolving each vector into its x and y components.

D. A car weighing $3,\overline{0}00$ lb is parked in a driveway that makes an angle of 8° with the horizontal. Find the minimum brake force that is needed to keep the car from rolling down the driveway. (Refer to the illustration in the text for Exercise 24.)

Exercise 20 See text for question.

Solution The following table organizes the solution.

Vector	Magnitude	Direction	x Component	y Component
A	5.0	60°	$5.0 \cos 60° = 2.500$	$5.0 \sin 60° = 4.330$
B	2.0	210°	$2.0 \cos 210° = -1.732$	$2.0 \sin 210° = -1.000$
C	1.0	270°	$1.0 \cos 270° = 0$	$1.0 \sin 270° = -1.000$
R			$R_x = 0.768$	$R_y = 2.330$

The magnitude of the resultant is

$$R = \sqrt{(R_x)^2 + (R_y)^2} = \sqrt{(0.768)^2 + (2.330)^2} = 2.5 \text{ lb}.$$

Since the x and y components are positive, θ_R is in Q_1. Then

$$\tan \theta_R = \frac{R_y}{R_x} = \frac{2.330}{0.768} = 3.0340 \text{ so } \theta_R = 72°.$$

Now do **C**.

Exercise 24 See text for question.

Solution Consider the illustration in the text. Let vector **OA** drawn vertically down represent the weight of the car, 3,500 lb. The resolution of vector **OA** into components that are parallel and perpendicular to the inclined plane results in vector **OB** (the force tending to pull the object down the plane) and vector **OC** (the pressure of the car on the driveway). The vector that we seek is vector **OD**, which has the same magnitude, but the opposite direction, of vector **OB**.

We can find the magnitude of **OB** by using the cosine function in right triangle *OBA*. Notice that vector **OA** makes a right angle with the horizontal. Thus, if the plane is inclined 5°, angle *BOA* is 85°. Then

$$\cos 85° = \overline{OB}/3,500 \quad \text{so} \quad \overline{OB} = 3,500 \cos 85°$$
$$= 310 \text{ lb (two significant digits)}.$$

Thus, a minimum brake force of 310 lb is required.
Now do **D**.

7.4 Trigonometric form of complex numbers

Objectives checklist

Can you:

a. Graph a complex number $a + bi$?

b. Write a complex number $a + bi$ in trigonometric form, and vice versa?

c. Find the product and quotient of complex numbers in trigonometric form?

d. Use De Moivre's theorem to find powers and roots of a complex number?

Key terms

Complex plane
Real axis
Imaginary axis

Argument of a complex number
Absolute value of a complex number

Key rules and formulas

- The complex number $a + bi$ is represented in a rectangular coordinate system by the point (a,b) with x value a and y value b.

- In trigonometric form for a complex number we write

$$a + bi = r(\cos \theta + i \sin \theta),$$

where the absolute value, r, is given by

$$r = \sqrt{a^2 + b^2}$$

and the argument, θ, is given by

$$\tan \theta = \frac{b}{a}.$$

- The product of complex numbers in trigonometric form is

$$r_1(\cos \theta_1 + i \sin \theta_1) \cdot r_2 (\cos \theta_2 + i \sin \theta_2)$$
$$= r_1 r_2[\cos(\theta_1 + \theta_2) + i \sin (\theta_1 + \theta_2)].$$

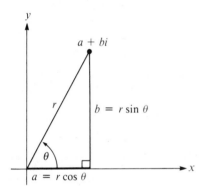

- The quotient of complex numbers in trigonometric form with $r_2 \neq 0$ is

$$\frac{r_1(\cos \theta_1 + i \sin \theta_1)}{r_2(\cos \theta_2 + i \sin \theta_2)} = \frac{r_1}{r_2}[\cos(\theta_1 - \theta_2) + i \sin(\theta_1 - \theta_2)].$$

- *De Moivre's theorem:* If $r(\cos \theta + i \sin \theta)$ is any complex number and n is any real number, then

$$[r(\cos \theta + i \sin \theta)]^n = r^n(\cos n\theta + i \sin n\theta).$$

- Any complex number has two square roots, three cube roots, four fourth roots, and so on. To find these roots, we use De Moivre's theorem to find one root, add 360° to θ, use the new trigonometric representation to find another root, and repeat this procedure until all n roots are found.

Additional comments

- The expression $r(\cos \theta + i \sin \theta)$ is often abbreviated as

$$r \text{ cis } \theta \quad \text{or} \quad r < \theta.$$

Detailed solutions to selected exercises

Exercise 6 Write $-1 + i$ in trigonometric form.

7.4

A. Write $\sqrt{2} + \sqrt{2}i$ in trigonometric form.

Solution The complex number $-1 + i$ is represented in a rectangular coordinate system by the point $(-1,1)$. To write $-1 + i$ in trigonometric form, we first determine the absolute value, r.

$$r = \sqrt{a^2 + b^2} = \sqrt{(-1)^2 + 1^2} = \sqrt{2}$$

Now determine the argument, θ.

$$\tan \theta = \frac{b}{a} = \frac{1}{-1} = -1.$$

Since $(-1,1)$ is in Q_2 and the reference angle is $45°$, we conclude that $\theta = 180° -$ reference angle $= 180° - 45° = 135°$. Thus,

$$-1 + i = \sqrt{2}(\cos 135° + i \sin 135°).$$

Now do **A.**

B. Write the number $2\sqrt{2}(\cos 45° + i \sin 45°)$ in the form $a + bi$.

Exercise 20 Write the number $2(\cos 210° + i \sin 210°)$ in the form $a + bi$.

Solution Since $\cos 210° = -\sqrt{3}/2$ and $\sin 210° = -1/2$, we have

$$2(\cos 210° + i \sin 210°) = 2\left(-\frac{\sqrt{3}}{2} + i\left(-\frac{1}{2}\right)\right) = -\sqrt{3} - i.$$

Now do **B.**

C. If $z_1 = 3(\cos 135° + i \sin 135°)$ and $z_2 = \cos 90° + i \sin 90°$, find (a) $z_1 \cdot z_2$, (b) z_1/z_2, (c) z_2/z_1.

Exercise 32 If $z_1 = 3\sqrt{2}(\cos 315° + i \sin 315°)$ and $z_2 = \cos 0° + i \sin 0°$, then find (a) $z_1 \cdot z_2$, (b) z_1/z_2, and (c) z_2/z_1.

Solution Using the product and quotient formulas, we have

a. $z_1 \cdot z_2 = (3\sqrt{2} \cdot 1)[\cos(315° + 0°) + i \sin(315° + 0°)]$
$= 3\sqrt{2}(\cos 315° + i \sin 315°)$

b. $z_1/z_2 = (3\sqrt{2} \div 1)[\cos(315° - 0°) + i \sin(315° - 0°)]$
$= 3\sqrt{2}(\cos 315° + i \sin 315°)$

c. $z_2/z_1 = (1 \div 3\sqrt{2})[\cos(0° - 315°) + i \sin(0° - 315°)]$
$= (\sqrt{2}/6)[\cos(-315°) + i \sin(-315°)]$

Note in part c that $1 \div 3\sqrt{2} = (1 \cdot \sqrt{2})/(3\sqrt{2} \cdot \sqrt{2}) = \sqrt{2}/6$.
Now do **C.**

D. Find $(-2 - 2i)^6$.

Exercise 36 Find $(-1 + i)^6$ and express the result in the form $a + bi$.

Solution As shown in the solution to Exercise 6,

$$-1 + i = \sqrt{2}(\cos 135° + i \sin 135°).$$

Then by De Moivre's theorem, we have

$$[\sqrt{2}(\cos 135° + i \sin 135°)]^6$$
$$= (\sqrt{2})^6[\cos(6 \cdot 135°) + i \sin(6 \cdot 135°)]$$
$$= 8(\cos 810° + i \sin 810°).$$

Since $\cos 810° = 0$ and $\sin 810° = 1$, $(-1 + i)^6 = 0 + 8i$.
Now do **D.**

Exercise 48 Solve $x^3 - 27i = 0$.

Solution $x^3 - 27i = 0$, so $x^3 = 27i$, so $x = (27i)^{1/3}$. In trigonometric form

$$27i = 27(\cos 90° + i \sin 90°).$$

Then we find the three cube roots of 27i by De Moivre's theorem.

E. Solve $x^3 - i = 0$.

$$\text{1st root} = [27(\cos 90° + i \sin 90°)]^{1/3}$$
$$= 27^{1/3}[\cos(90°/3) + i \sin(90°/3)]$$
$$= 3(\cos 30° + i \sin 30°)$$
$$= \frac{3\sqrt{3}}{2} + \frac{3}{2}i$$

$$\text{2nd root} = 27^{1/3}\left(\cos \frac{90° + 360°}{3} + i \sin \frac{90° + 360°}{3}\right)$$
$$= 3(\cos 150° + i \sin 150°)$$
$$= -\frac{3\sqrt{3}}{2} + \frac{3}{2}i$$

$$\text{3rd root} = 27^{1/3}\left(\cos \frac{90° + 2 \cdot 360°}{3} + i \sin \frac{90° + 2 \cdot 360°}{3}\right)$$
$$= 3(\cos 270° + i \sin 270°)$$
$$= 0 - 3i.$$

Now do **E.**

7.5 Polar coordinates

Objectives checklist

Can you:

a. Plot points in polar coordinates?

b. Give different polar representations for a given point?

c. Classify and graph certain polar equations associated with circles, cardioids, and roses?

d. Graph an equation in polar coordinates by plotting points and testing for x- and y-axis symmetry?

e. Convert a point from rectangular coordinates to polar coordinates, and vice versa?

f. Transform an equation from rectangular form to polar form, and vice versa?

Key Terms

Polar coordinate system
Polar coordinates
Pole (or origin)

Polar axis
Cardioid
Rose

Key rules and formulas

- If $r > 0$, we plot (r,θ) by measuring r units along the terminal ray of θ. If $r < 0$, we plot (r,θ) by measuring $|r|$ units in a direction opposite to the terminal ray of θ. If $r = 0$, (r,θ) is the pole for all values of θ.

- Different representations for a given point are obtained as follows:

$$(r,\theta) = (r,\theta + n \cdot 360°), \text{ where } n \text{ is an integer}$$
$$(-r,\theta) = (r,\theta + n \cdot 180°), \text{ where } n \text{ is an odd integer.}$$

- A curve has x-axis symmetry if replacing θ by $-\theta$ in its equation produces an equivalent equation. A curve has y-axis symmetry if replacing θ by $180 - \theta$ in its equation produces an equivalent equation.

• Graphs of certain circles may be identified from their equations as follows:

Equation	Description of Graph
$r = a$	A circle of radius a and centered at the origin
$r = 2a \sin \theta$	A circle of radius a, centered on the positive y-axis, and passing through the origin
$r = -2a \sin \theta$	A circle of radius a, centered on the negative y-axis, and passing through the origin
$r = 2a \cos \theta$	A circle of radius a, centered on the positive x-axis, and passing through the origin
$r = -2a \cos \theta$	A circle of radius a, centered on the negative x-axis, and passing through the origin

• Equations of the form $r = a + a \sin \theta$, $r = a - a \sin \theta$, $r = a + a \cos \theta$, and $r = a - a \cos \theta$ graph as cardioids. The sine versions are symmetric with respect to the y-axis, and the cosine versions are symmetric with respect to the x-axis.

• Equations of the form $r = a \sin n\theta$ and $r = a \cos n\theta$ graph as roses. The rose has n loops when n is odd and $2n$ loops when n is even. The values of θ for which $|r|$ is greatest yield endpoints for the loops.

• Relations between polar and rectangular coordinates:

$$y = r \sin \theta$$
$$x = r \cos \theta$$
$$\tan \theta = \frac{y}{x}$$
$$x^2 + y^2 = r^2.$$

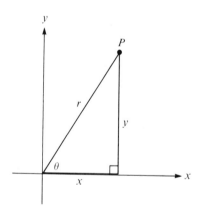

Detailed solutions to selected exercises

Exercise 2d Plot $(-2, 210°)$ in polar coordinates.

Solution We are given that $r = -2$ and $\theta = 210°$. Since r is negative, the point is located 2 units in the opposite direction from the terminal ray of a 210° angle as shown in Figure 7.5:2d.

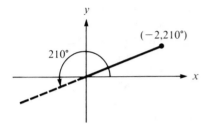

Figure 7.5:2d

Now do **A**.

Exercise 4 List three other polar representations for (3,210°), with one having $r < 0$.

Solution To obtain alternate polar representations to (r,θ), we change θ by 360° and keep r the same or change θ by 180° and change the sign on r. Thus,

$$(3,210°) = (3,210° + 360°) \quad = (3,570°)$$
$$\text{or } (3,210°) = (3,210° - 360°) \quad = (3,-150°)$$
$$\text{or } (3,210°) = (-3,210° - 180°) = (-3,30°).$$

Now do **B**.

Exercise 14 Graph $r = 1 + \cos \theta$ in polar coordinates and give its name.

Solution The equation fits the form $r = a + a \cos \theta$, so its graph is a cardioid that is symmetric with respect to the x-axis. We can determine where the graph intersects the axes by substituting the values of 0°, 90°, 180°, and 270° for θ in $r = 1 + \cos \theta$ to obtain the points (2,0°), (1,90°), (0,180°), and (1,270°). Plotting these points and using our knowledge about the shape of the cardioid and its symmetry gives the graph in Figure 7.5:14.

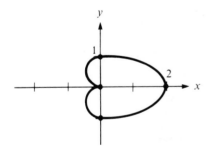

Figure 7.5:14

Now do **C**.

Exercise 24 Graph $r = 1 + 2 \cos \theta$ in polar coordinates (limacon, inner loop).

Solution Since the equation does not match the form of a circle, cardioid, or rose, we graph it by plotting points. Fewer points will be needed if it is first determined that the curve has x- or y-axis symmetry. Testing for y-axis symmetry by replacing θ by 180° $- \theta$ in the equation $r = 1 + 2 \cos \theta$ yields

7.5

A. Plot $(-1,150°)$ in polar coordinates.

B. List three other polar representations for (2,135°) with one having $r < 0$.

C. Graph $r = 1 + \sin \theta$ and give its name.

128

D. Graph $r = 1 - 2 \sin \theta$ (limacon, inner loop).

E. Express $(2,\pi/4)$ in rectangular coordinates.

F. Express $(-3,-3\sqrt{3})$ in polar coordinates with $r \geq 0$ and $0° \leq \theta < 360°$.

$r = 1 + 2 \cos(180° - \theta)$. Since $\cos(180° - \theta) = -\cos \theta$, we have $r = 1 + 2(-\cos \theta) = 1 - 2 \cos \theta$, which is not equivalent to the original equation. Thus, we have not found y-axis symmetry. Testing for x-axis symmetry by replacing θ by $-\theta$ in $r = 1 + 2 \cos \theta$ yields $r = 1 + 2 \cos(-\theta)$, which is equivalent to $r = 1 + 2 \cos \theta$, since $\cos(-\theta) = \cos \theta$. Thus, we have found x-axis symmetry. Now we can graph half of the curve and then reflect this about the x-axis to complete the graph. Picking some values of θ ranging from $0°$ to $180°$ (say $0°$, $30°$, $60°$, $90°$, $120°$, $150°$, and $180°$) and substituting them into $r = 1 + 2 \cos \theta$ generates the ordered pairs $(0°,3)$, $(30°,2.732)$, $(60°,2)$, $(90°,1)$, $(120°,0)$, $(150°,-1.732)$, and $(180°,-1)$. Plotting these points, connecting them, and then reflecting the curve about the x-axis results in the graph shown in Figure 7.5:24.

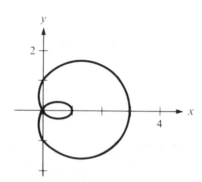

Figure 7.5:24

Now do **D.**

Exercise 34 Express $(-\sqrt{3},-120°)$ in rectangular coordinates.

Solution Using the formulas $x = r \cos \theta$ and $y = r \sin \theta$ with $r = -\sqrt{3}$ and $\theta = -120°$ yields

$$x = -\sqrt{3} \cos(-120°) = -\sqrt{3}(-\cos 60°) = -\sqrt{3}(-\tfrac{1}{2}) = \sqrt{3}/2$$
$$y = -\sqrt{3} \sin(-120°) = -\sqrt{3}(-\sin 60°) = -\sqrt{3}(-\sqrt{3}/2) = 3/2.$$

Thus, $(-\sqrt{3},-120°)$ is written as $(\sqrt{3}/2,\tfrac{3}{2})$ in rectangular coordinates. Now do **E.**

Exercise 38 Express $(-3,-3)$ in polar coordinates with $r \geq 0$ and $0° \leq \theta < 360°$.

Solution From $r^2 = x^2 + y^2$ with $x = -3$ and $y = -3$, we obtain

$$r^2 = (-3)^2 + (-3)^2, \text{ so } r^2 = 18, \text{ so } r = \sqrt{18} = 3\sqrt{2}.$$

Also $\tan \theta = y/x = (-3)/(-3) = 1$. Since $(-3,-3)$ is in Q_3 and $r \geq 0$, we conclude $\theta = 225°$. Thus, $(-3,-3)$ is written in polar coordinates as $(3\sqrt{2},225°)$.
Now do **F.**

Exercise 48 Transform $y = x^3$ to polar form.

Solution Using the relations $x = r \cos \theta$ and $y = r \sin \theta$, we obtain

$r \sin \theta = r^3 \cos^3 \theta$, so $r^3 \cos^3 \theta - r \sin \theta = 0$, so $r(r^2 \cos^3 \theta - \sin \theta) = 0$.

Setting each factor equal to 0 gives $r = 0$ or $r^2 = \sin \theta / \cos^3 \theta$. Since $r^2 = \sin \theta / \cos^3 \theta$ includes the pole, or $r = 0$, the complete relation is expressed by $r^2 = \sin \theta / \cos^3 \theta$. We may also write this equation as

$$r^2 = \frac{\sin \theta}{\cos \theta} \cdot \frac{1}{\cos^2 \theta} = \tan \theta \sec^2 \theta.$$

Now do **G**.

Exercise 56 Transform $r = \sec \theta$ to rectangular form.

Solution Since $\sec \theta = 1/\cos \theta$, we have

$$r = \sec \theta = 1/\cos \theta, \text{ so } r \cos \theta = 1.$$

Then $r \cos \theta = x$, so the equation in rectangular form is $x = 1$.
Now do **H**.

G. Transform $x = y^2$ to polar form.

H. Transform $r = \csc \theta$ to rectangular form.

Sample test questions: Chapter 7

1. In triangle ABC, $A = 70°$, $B = 80°$, and $a = 12$ m. Find c.

2. A body is acted on by two forces with magnitudes of 4.0 lb and 9.0 lb, which act at an angle of 45° with each other. Find the angle between the resultant and the larger force.

3. If a vector with magnitude 11 lb makes an angle of 30° with the positive x-axis, find the vertical component of the vector.

4. Write the following numbers in trigonometric form.

   a. $1 - i$ b. $-\sqrt{2} - \sqrt{2}i$ c. i

5. Express the point $(3\sqrt{3}, 3)$ in polar coordinates with $r \geq 0$ and $0° \leq \theta < 360°$.

6. Express the point $(2, 5\pi/6)$ in rectangular coordinates.

7. Transform the equation $x^2 + y^2 + x = \sqrt{x^2 + y^2}$ to polar form.

8. Graph each equation in polar coordinates.

   a. $r = \cos \theta$ b. $r = \sin \theta - 1$

9. Select the choice that completes the statement or answers the question.

   a. To solve a triangle when given the lengths of the three sides, we use the

     (a) Pythagorean theorem   (c) law of cosines

     (b) law of sines   (d) tangent function

   b. In triangle ABC, if $a = 6.0$, $b = 9.0$, and $c = 7.0$, then B equals

     (a) 74°   (b) 79°   (c) 82°   (d) 87°

c. An alternate polar representation for $(-2, 60°)$ is

   (a) $(2, 60°)$ (b) $(2, 120°)$ (c) $(-2, 120°)$ (d) $(2, 240°)$

d. In the form $a + bi$ we write $-3(\cos 135° + i \sin 135°)$ as

   (a) $(3/\sqrt{2}) - (3/\sqrt{2})i$ (b) $(-3/\sqrt{2}) - (3/\sqrt{2})i$

   (c) $(3/\sqrt{2}) + (3/\sqrt{2})i$ (d) $3 - 3i$

e. $(1 - i)^4$ simplifies to

   (a) 4 (b) -4 (c) $4i$ (d) $-4i$

f. In triangle ABC, if $\sin A = \frac{1}{2}$ and $\sin B = \frac{5}{6}$, then the ratio of side length a to side length b is

   (a) $3:5$ (b) $6:5$ (c) $4:3$ (d) $1:5$

g. The graph of $r = 2 \sin 3\theta$ is a

   (a) circle   (b) cardioid   (c) rose   (d) line

Chapter 8
Analytic Geometry: Conic Sections

8.1 Introduction to analytic geometry

Objectives checklist

Can you:

a. Find an equation for a graph that is defined by certain geometric conditions?

b. State the fundamental relationship between an equation and its graph?

c. Find the midpoint of the line segment joining a given pair of points?

d. Find an equation for the line that is the perpendicular bisector of the line segment joining two given points?

e. Use the methods of analytic geometry to prove geometric theorems?

Key rules and formulas

- The fundamental relationship between an equation and its graph is as follows:

  Every ordered pair that satisfies the equation corresponds to a point in its graph, and every point in the graph corresponds to an ordered pair that satisfies the equation.

- The given geometric conditions in this section involve distances so the following distance formulas are needed.

  $$d = |x_2 - x_1| \quad \text{(for two points with the same } y\text{-coordinate)}$$
  $$d = |y_2 - y_1| \quad \text{(for two points with the same } x\text{-coordinate)}$$
  $$d = \sqrt{(x_2 - x_1)^2 + (y_2 - y_1)^2} \quad \text{(for any two points)}$$

- *Midpoint formula:* The midpoint of the line segment joining (x_1, y_1) and (x_2, y_2) is

  $$\left(\frac{x_1 + x_2}{2}, \frac{y_1 + y_2}{2} \right).$$

- To prove a geometric theorem using analytic geometry, we introduce a coordinate system and use algebraic methods. The figure specified in the theorem may be placed in any convenient position on the coordinate system. Frequently we place a vertex of the figure at the origin and at least one side of the figure on one of the coordinate axes.

Detailed solutions to selected exercises

Exercise 2 Find an equation for the graph that is the set of all points in a plane equidistant from $(1, -3)$ and $(-4, 2)$.

Solution Let $P(x, y)$ represent any point satisfying the given condition. Then since (x, y) is equidistant from the given points, we have

$$\sqrt{(x - 1)^2 + [y - (-3)]^2} = \sqrt{[x - (-4)]^2 + (y - 2)^2}.$$

Chapter 8

8.1

A. Find an equation for the graph that is the set of all points in a plane equidistant from $(-5,1)$ and $(3,2)$.

Squaring both sides and expanding gives

$$x^2 - 2x + 1 + y^2 + 6y + 9 = x^2 + 8x + 16 + y^2 - 4y + 4$$
$$0 = 10x - 10y + 10$$
$$0 = x - y + 1.$$

The graph of $0 = x - y + 1$ satisfies the given geometric condition.
Now do **A.**

Exercise 10 Find an equation for the graph that is the set of all points in a plane the difference of whose distances from $(4,0)$ and $(-4,0)$ is 6.

Solution For any point $P(x,y)$ on the graph we have

$$d_1 = \sqrt{(x - 4)^2 + (y - 0)^2} \quad \text{and} \quad d_2 = \sqrt{[x - (-4)]^2 + (y - 0)^2}.$$

According to the given geometric condition,

$$|d_1 - d_2| = 6 \quad \text{so} \quad d_1 - d_2 = \pm 6.$$

Then

$$\sqrt{(x - 4)^2 + y^2} - \sqrt{(x + 4)^2 + y^2} = \pm 6$$
$$\sqrt{(x - 4)^2 + y^2} = \pm 6 + \sqrt{(x + 4)^2 + y^2}$$

B. Find an equation for the graph that is the set of all points in a plane the difference of whose distances from $(2,0)$ and $(-2,0)$ is 1.

Square both sides of the equation and simplify.

$$(x - 4)^2 + y^2 = 36 \pm 12\sqrt{(x + 4)^2 + y^2} + (x + 4)^2 + y^2$$
$$-16x - 36 = \pm 12\sqrt{(x + 4)^2 + y^2}$$
$$-4x - 9 = \pm 3\sqrt{(x + 4)^2 + y^2}$$
$$16x^2 + 72x + 81 = 9(x^2 + 8x + 16 + y^2)$$
$$0 = 9y^2 - 7x^2 + 63$$

The graph of $0 = 9y^2 - 7x^2 + 63$ satisfies the given condition.
Now do **B.**

Exercise 12 Find the midpoint of the line segment joining the points $(5,-4)$ and $(-9,0)$.

C. Find the midpoint of the line segment joining the points $(-3,8)$ and $(9,-4)$.

Solution Using the midpoint formula with $x_1 = 5$, $y_1 = -4$, $x_2 = -9$, and $y_2 = 0$ gives

$$\left(\frac{5 + (-9)}{2}, \frac{-4 + 0}{2} \right).$$

Thus, the midpoint of the line segment is the point $(-2,-2)$.
Now do **C.**

Exercise 22 Find an equation for the line that is the perpendicular bisector of the line segment with endpoints $(1,0)$ and $(-2,7)$.

Solution First, the midpoint of the given line segment is $(-\frac{1}{2}, \frac{7}{2})$ since

$$\left(\frac{x_1 + x_2}{2}, \frac{y_1 + y_2}{2} \right) = \left(\frac{1 + (-2)}{2}, \frac{0 + 7}{2} \right) = \left(-\frac{1}{2}, \frac{7}{2} \right).$$

Also, the slope of the given line segment is

$$m = \frac{y_2 - y_1}{x_2 - x_1} = \frac{7 - 0}{-2 - 1} = -\frac{7}{3}.$$

Thus, the perpendicular bisector passes through $(-\frac{1}{2},\frac{7}{2})$ and has slope $\frac{3}{7}$ (the negative reciprocal of $-\frac{7}{3}$). From the point-slope equation, the perpendicular bisector is

$$y - \tfrac{7}{2} = \tfrac{3}{7}[x - (-\tfrac{1}{2})] \text{ so } y - \tfrac{7}{2} = \tfrac{3}{7}x + \tfrac{3}{14} \text{ so } y = \tfrac{3}{7}x + \tfrac{26}{7}.$$

Now do **D**.

D. Find an equation for the line that is the perpendicular bisector of the line segment with endpoints $(-6,-2)$ and $(0,8)$.

Exercise 28 Use the figure given in the text to prove that the midpoints of the sides of a rectangle are the vertices of a rhombus (that is, a parallelogram with all sides equal in length).

Solution By the midpoint formula, the midpoint, D, of AB is $(a,b/2)$, the midpoint, E, of BC is $(a/2,b)$, the midpoint, F, of OC is $(0,b/2)$, and the midpoint, G, of OA is $(a/2,0)$. Consider the quadrilateral having these points as vertices. We determine the lengths of sides DE, EF, FG, and GD by using the formula for the distance between two points, which gives

$$\overline{DE} = \sqrt{\left(a - \frac{a}{2}\right)^2 + \left(\frac{b}{2} - b\right)^2} = \sqrt{\frac{a^2}{4} + \frac{b^2}{4}}$$

$$\overline{EF} = \sqrt{\left(\frac{a}{2} - 0\right)^2 + \left(b - \frac{b}{2}\right)^2} = \sqrt{\frac{a^2}{4} + \frac{b^2}{4}}$$

$$\overline{FG} = \sqrt{\left(0 - \frac{a}{2}\right)^2 + \left(\frac{b}{2} - 0\right)^2} = \sqrt{\frac{a^2}{4} + \frac{b^2}{4}}$$

$$\overline{GD} = \sqrt{\left(\frac{a}{2} - a\right)^2 + \left(0 - \frac{b}{2}\right)^2} = \sqrt{\frac{a^2}{4} + \frac{b^2}{4}}.$$

Thus, all the sides are the same length. Also, by the slope formula,

$$m_{DE} = m_{FG} = -\frac{b}{a}, \text{ while } m_{EF} = m_{GD} = \frac{b}{a}.$$

So DE and FG are parallel and EF and GD are parallel. Thus, the quadrilateral with vertices D, E, F, and G is a rhombus.

8.2 The circle

Objectives checklist

Can you:

a. State the manner in which a plane and cone must intersect for the intersection to be a circle, an ellipse, a hyperbola, or a parabola?

b. Find the equation of a circle in standard form given its center and radius?

c. Find the equation of a circle in standard form given its center and either a point on the circle or a vertical or horizontal tangent line to the curve?

d. Find the center and radius of a circle given in the standard form $(x - h)^2 + (y - k)^2 = r^2$?

e. Find the center and radius of a circle that is not in standard form by completing the square?

Key terms

Conic sections
Circle
Parabola

Ellipse
Hyperbola

8.2

A. Find the equation in standard form of the circle with center $(-1,5)$ that is tangent to the line $y = 3$.

Key rules and formulas

- The intersection of a plane and a cone produces the four major conic sections. If the cutting plane does not contain the vertex, then the important conic sections are obtained as follows:

  1. A circle: The intersecting plane is parallel to the base of the cone.
  2. A parabola: The intersecting plane is parallel to the side of the cone.
  3. A hyperbola: The intersecting plane cuts both nappes of the cone.
  4. An ellipse: The intersecting plane is parallel to neither the base nor the side and intersects only one nappe of the cone.

  These cases are illustrated in text Figure 8.7. Three degenerate cases occur when the cutting plane passes through the vertex. These degenerate conic sections are a point, a line, or a pair of intersecting lines.

- The equation in standard form of a circle of radius r with center (h,k) is

$$(x - h)^2 + (y - k)^2 = r^2.$$

- We complete the square for $x^2 + bx$ by adding $(b/2)^2$ which is the square of one half of the coefficient of x.

Detailed solutions to selected exercises

Exercise 6 Find the equation in standard form of the circle with center at $(1,4)$ that is tangent to the line $y = -3$.

Solution Substituting $h = 1$ and $k = 4$ into $(x - h)^2 + (y - k)^2 = r^2$ gives

$$(x - 1)^2 + (y - 4)^2 = r^2.$$

The tangent line $y = -3$ intersects the circle in exactly one point with the tangent being perpendicular to the radius at the point of intersection. As shown in Figure 8.2:6, the intersection point is $(1,-3)$ and the radius of the circle is 7. Since $r = 7$, $r^2 = 49$ and the standard equation of the circle is $(x - 1)^2 + (y - 4)^2 = 49$.

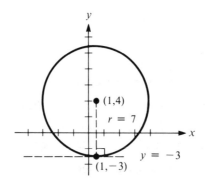

Figure 8.2:6

Now do **A.**

Exercise 14 Find the center and radius of $x^2 + (y + 2)^2 = 5$.

Solution To match standard form, the given equation may be rewritten as

$$(x - 0)^2 + [y - (-2)]^2 = (\sqrt{5})^2.$$

Thus, the center is $(0,-2)$ and the radius is $\sqrt{5}$.
Now do **B.**

Exercise 22 Find the center and radius of $x^2 + y^2 - 5x - y - 3 = 0$.

Solution We complete the square for both the x terms and the y terms to obtain the standard form as follows:

$$(x^2 - 5x) + (y^2 - y) = 3$$
$$(x^2 - 5x + \tfrac{25}{4}) + (y^2 - y + \tfrac{1}{4}) = 3 + \tfrac{25}{4} + \tfrac{1}{4}$$
$$(x - \tfrac{5}{2})^2 + (y - \tfrac{1}{2})^2 = \tfrac{38}{4}.$$

Thus, the center is $(\tfrac{5}{2}, \tfrac{1}{2})$, and $r^2 = \tfrac{38}{4}$ so the radius is $\sqrt{38}/2$.
Now do **C.**

B. Find the center and radius of $(x + 7)^2 + y^2 = 17$.

C. Find the center and radius of $x^2 + y^2 - x - 2y - 2 = 0$.

8.3 The ellipse

Objectives checklist

Can you:

a. Find the coordinates of the center, the foci, and the endpoints of the major and minor axes given an equation of an ellipse?

b. Find an equation for an ellipse satisfying various conditions concerning its center, foci, vertices, and major and minor axes?

c. Graph an ellipse?

Key terms

Ellipse   Major axis
Foci   Minor axis
Vertices

Key rules and formulas

*   The standard form of the equation of an ellipse with center at the origin and foci on the x-axis is

$$\frac{x^2}{a^2} + \frac{y^2}{b^2} = 1.$$

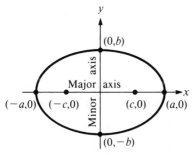

Foci: $(-c,0)$, $(c,0)$
Vertices: $(-a,0)$, $(a,0)$

- The standard form of the equation of an ellipse with center at the origin and foci on the y-axis is

$$\frac{x^2}{b^2} + \frac{y^2}{a^2} = 1.$$

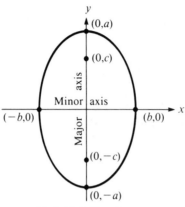

Foci: $(0,c)$, $(0,-c)$
Vertices: $(0,a)$, $(0,-a)$

- In the standard form of the equations of an ellipse,

    a is the distance from the center to the endpoints on the major axis
    b is the distance from the center to the endpoints on the minor axis
    c is the distance from the center to the foci.

    These distances are related by the formula

$$a^2 = b^2 + c^2.$$

- If an ellipse is centered at the point (h,k), and if the major axis is parallel to the x-axis, then the standard form of the equation is

$$\frac{(x - h)^2}{a^2} + \frac{(y - k)^2}{b^2} = 1.$$

    If the major axis is parallel to the y-axis, the standard form of the equation is

$$\frac{(x - h)^2}{b^2} + \frac{(y - k)^2}{a^2} = 1.$$

Detailed solutions to selected exercises

Exercise 4 For $(x^2/49) + (y^2/9) = 1$, find the coordinates of the foci and the endpoints of the major and minor axes. Also, sketch the ellipse.

Solution The equation is in standard form in the case where the larger denominator, a^2, is in the x term so the major axis is along the x-axis. Then

$$a^2 = 49, \text{ so } a = 7 \quad \text{and} \quad b^2 = 9, \text{ so } b = 3.$$

To find c, substitute these values in the formula $a^2 = b^2 + c^2$.

$$49 = 9 + c^2 \quad \text{so} \quad c^2 = 40, \quad \text{so} \quad c = \sqrt{40}.$$

Now using these values for a, b, and c, we have

endpoints on major axis: $(7,0)$, $(-7,0)$
endpoints on minor axis: $(0,3)$, $(0,-3)$
coordinates of foci: $(\sqrt{40},0)$, $(-\sqrt{40},0)$.

The ellipse is graphed in Figure 8.3:4.

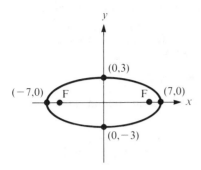

Figure 8.3:4

Now do **A.**

Exercise 14   Find the coordinates of the foci and the endpoints of the major and minor axes for $9x^2 + y^2 - 36x + 2y + 1 = 0$. Also, sketch the ellipse.

Solution   First, put the equation in standard form by completing the square.

$$(9x^2 - 36x \quad) + (y^2 + 2y \quad) = -1$$
$$9(x^2 - 4x \quad) + (y^2 + 2y \quad) = -1$$
$$9(x^2 - 4x + 4) + (y^2 + 2y + 1) = -1 + 36 + 1$$
$$\frac{(x-2)^2}{4} + \frac{(y+1)^2}{36} = 1.$$

From this equation in standard form we conclude the center is at $(2,-1)$. Also, since the larger denominator is in the y term, the major axis is parallel to the y-axis, and

$$a^2 = 36, \text{ so } a = 6 \quad \text{and} \quad b^2 = 4, \text{ so } b = 2.$$

To find c, replace a^2 by 36 and b^2 by 4 in the formula $a^2 = b^2 + c^2$.

$$36 = 4 + c^2, \text{ so } c^2 = 32 \quad \text{and} \quad c = \sqrt{32}.$$

Now using these values for a, b, and c, we have

endpoints on major axis: $(2,-1 + 6), (2,-1 - 6) = (2,5), (2,-7)$
endpoints on minor axis: $(2 + 2,-1), (2 - 2,-1) = (4,-1), (0,-1)$
coordinates of foci: $(2,-1 + \sqrt{32}), (2,-1 - \sqrt{32})$.

The ellipse is graphed in Figure 8.3:14.
Now do **B.**

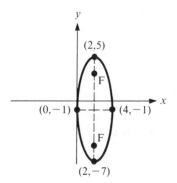

Figure 8.3:14

8.3

A. For $(x^2/25) + (y^2/16) = 1$, find the coordinates of the foci and the endpoints of the major and minor axes. Also, sketch the ellipse.

B. Find the coordinates of the foci and the endpoints of the major and minor axes for the graph of $4x^2 + 25y^2 - 24x + 50y = 39$. Also, sketch the ellipse.

137

C. Find an equation in standard form for the ellipse with vertices $(-3,3)$ and $(7,3)$ and focus at $(6,3)$.

D. An elliptical arch has a height of 40 ft and a span of 100 ft. How high is the arch 30 ft each side of center?

Exercise 24   Find an equation in standard form for the ellipse with vertices at $(4,1)$ and $(-8,1)$ and focus at $(-4,1)$.

Solution   The center is at the midpoint of the major axis, so the center is $(-2,1)$. The distance from the center $(-2,1)$ to a vertex, say $(4,1)$, is 6, so $a = 6$. Since c is the distance from the center to a focus, c is 2. To find b^2, replace a by 6 and c by 2 in the formula $a^2 = b^2 + c^2$.

$$36 = b^2 + 4 \quad \text{so} \quad b^2 = 32$$

Finally, the major axis is parallel to the x-axis, so the equation in standard form

$$\frac{(x-h)^2}{a^2} + \frac{(y-k)^2}{b^2} = 1 \quad \text{yields} \quad \frac{(x+2)^2}{36} + \frac{(y-1)^2}{32} = 1.$$

Now do **C.**

Exercise 26   An elliptical arch has a height of 20 ft and a span of 50 ft. How high is the arch 15 ft each side of center?

Solution   First, sketch Figure 8.3:26. Then the equation for the ellipse is of the form $(x^2/a^2) + (y^2/b^2) = 1$, where $a = 25$ and $b = 20$. Thus, the equation is $(x^2/625) + (y^2/400) = 1$. Now we find the height of the arch at $x = 15$ ft as follows:

$$\frac{15^2}{625} + \frac{y^2}{400} = 1$$

$$y^2 = 400\left(1 - \frac{225}{625}\right)$$

$$y = \sqrt{256} = 16 \text{ ft.}$$

The arch is 16 ft high at the specified point.

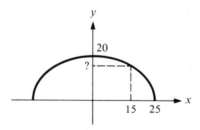

Figure 8.3:26

Now do **D.**

8.4 The hyperbola

Objectives checklist

Can you:

a. Find the coordinates of the center, the vertices, and the foci given an equation of a hyperbola?

b. Determine the asymptotes of a hyperbola given its equation?

c. Find an equation in standard form for a hyperbola satisfying various conditions concerning its center, foci, vertices, and transverse and conjugate axes?

d. Graph a hyperbola?

Key terms

Hyperbola	Transverse axis
Foci	Conjugate axis
Vertices	Asymptotes (of the hyperbola)

Key rules and formulas

• The standard form of the equation of a hyperbola with center at the origin and foci on the x-axis is

$$\frac{x^2}{a^2} - \frac{y^2}{b^2} = 1$$

and the asymptotes in this case are $y = \pm(b/a)x$.

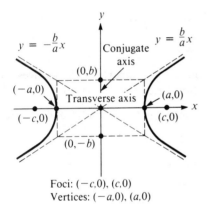

Foci: $(-c,0)$, $(c,0)$
Vertices: $(-a,0)$, $(a,0)$

• The standard form of the equation of a hyperbola with center at the origin and foci on the y-axis is

$$\frac{y^2}{a^2} - \frac{x^2}{b^2} = 1$$

and the asymptotes of the hyperbola in this case are $y = \pm(a/b)x$.

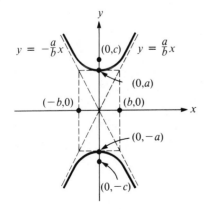

- In the standard form of the equation of a hyperbola,

  a is the distance from the center to the endpoints on the transverse axis
  b is the distance from the center to the endpoints on the conjugate axis
  c is the distance from the center to the foci.

  These distances are related by the formula

  $$c^2 = a^2 + b^2.$$

- If a hyperbola is centered at (h,k) and the transverse axis is parallel to the x-axis, then the standard form is

  $$\frac{(x - h)^2}{a^2} - \frac{(y - k)^2}{b^2} = 1$$

  and the asymptotes are given by $y - k = (\pm b/a)(x - h)$.

- If a hyperbola is centered at (h,k) and the transverse axis is parallel to the y-axis, then the standard form is

  $$\frac{(y - k)^2}{a^2} - \frac{(x - h)^2}{b^2} = 1$$

  and the asymptotes are given by $y - k = (\pm a/b)(x - h)$.

Detailed solutions to selected exercises

Exercise 4   For $(y^2/1) - (x^2/4) = 1$, find the coordinates of the vertices and the foci and determine the asymptotes. Also, sketch the hyperbola.

Solution   The equation is in the standard form $(y^2/a^2) - (x^2/b^2) = 1$. Then

$$a^2 = 1, \text{ so } a = 1 \quad \text{and} \quad b^2 = 4, \text{ so } b = 2.$$

To find c, substitute these values in the formula $c^2 = a^2 + b^2$.

$$c^2 = 1 + 4 = 5, \text{ so } c = \sqrt{5}.$$

Now since the transverse axis lies along the y-axis, we have

  vertices (endpoints on transverse axis): $(0,1)$, $(0,-1)$
  endpoints on conjugate axis: $(2,0)$, $(-2,0)$
  coordinates of foci: $(0,\sqrt{5})$, $(0,-\sqrt{5})$
  asymptotes: $y = \pm(a/b)x = \pm\frac{1}{2}x$.

The hyperbola is graphed in Figure 8.4:4.

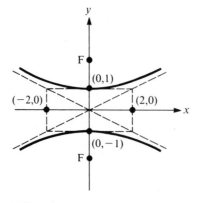

Figure 8.4:4

Now do **A.**

Exercise 14 Find the coordinates of the vertices and the foci and determine the asymptotes for $9y^2 - x^2 - 36y + 2x - 1 = 0$. Also, sketch the hyperbola.

Solution First, put the equation in standard form by completing the square.

$$(9y^2 - 36y) - (x^2 - 2x) = 1$$
$$9(y^2 - 4y) - (x^2 - 2x) = 1$$
$$9(y^2 - 4y + 4) - (x^2 - 2x + 1) = 1 + 9(4) + 1(-1)$$
$$9(y - 2)^2 - (x - 1)^2 = 36$$
$$\frac{(y - 2)^2}{4} - \frac{(x - 1)^2}{36} = 1$$

From this equation we conclude that the transverse axis is parallel to the y-axis, the center is at $(1,2)$, and

$$a^2 = 4, \text{ so } a = 2 \quad \text{and} \quad b^2 = 36, \text{ so } b = 6.$$

To find c, replace a^2 by 4 and b^2 by 36 in the formula $c^2 = a^2 + b^2$.

$$c^2 = 4 + 36 = 40, \text{ so } c = \sqrt{40}.$$

Now using these values for a, b, and c we have

vertices (endpoints on transverse axis): $(1, 2 + 2), (1, 2 - 2) = (1,4), (1,0)$
endpoints on conjugate axis: $(1 + 6, 2), (1 - 6, 2) = (7,2), (-5,2)$
coordinates of foci: $(1, 2 + \sqrt{40}), (1, 2 - \sqrt{40})$
asymptotes: $y - 2 = \pm\frac{1}{3}(x - 1)$.

The hyperbola is graphed in Figure 8.4:14.

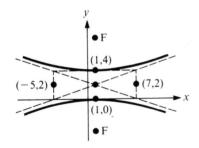

Figure 8.4:14

Now do **B.**

Exercise 24 Find an equation in standard form for the hyperbola with vertices at $(4,1)$ and $(-10,1)$ and focus at $(7,1)$.

Solution The center is at the midpoint of the transverse axis so the center is $(-3,1)$. The distance from the center to a vertex is 7, so $a = 7$; while the distance from the center to a focus at $(7,1)$ is 10, so $c = 10$. Then

$$c^2 = a^2 + b^2 \quad \text{so} \quad 100 = 49 + b^2 \quad \text{so} \quad b^2 = 51.$$

Finally, the transverse axis is parallel to the x-axis, so the standard form

$$\frac{(x - h)^2}{a^2} - \frac{(y - k)^2}{b^2} = 1 \text{ becomes } \frac{(x + 3)^2}{49} - \frac{(y - 1)^2}{51} = 1.$$

Now do **C.**

8.4

A. For $(x^2/36) - (y^2/4) = 1$, find the coordinates of the vertices and the foci and determine the asymptotes. Also, sketch the hyperbola.

B. Find the coordinates of the vertices and the foci and determine the asymptotes for $4y^2 - 25x^2 + 8y - 150x = 321$. Also, sketch the hyperbola.

C. Find an equation in standard form for the hyperbola with vertices at $(2,6)$ and $(-4,6)$ and focus at $(8,6)$.

8.5 The parabola

Objectives checklist

Can you:

a. Find the coordinates of the vertex and the focus, and the equation of the directrix, given an equation of a parabola?

b. Find an equation in standard form for a parabola satisfying various conditions concerning its vertex, focus, and directrix?

c. Graph a parabola?

Key terms

Parabola   Vertex
Focus   Axis of symmetry
Directrix

Key rules and formulas

• The graphs of the equations

$$y^2 = \pm 4px \quad \text{and} \quad x^2 = \pm 4py$$

are parabolas. In all cases p gives the distance from the vertex to the focus and from the vertex to the directrix.

• If the y term is squared, the axis of symmetry is the x-axis. The parabola opens to the right when the coefficient of x is positive and to the left if this coefficient is negative.

• If the x term is squared, the axis of symmetry is the y-axis. The parabola opens upward if the coefficient of y is positive and downward when this coefficient is negative.

• If the vertex of the parabola is at the point (h,k) and the directrix is parallel to the x- or y-axis, the four possible forms (with $p > 0$) for the parabola are shown here.

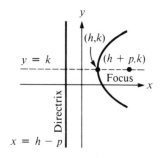

Case 1:
$(y - k)^2 = 4p(x - h)$

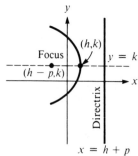

Case 2:
$(y - k)^2 = -4p(x - h)$

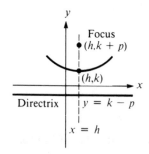

Case 3:
$(x - h)^2 = 4p(y - k)$

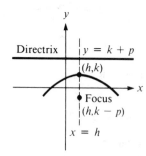

Case 4:
$(x - h)^2 = -4p(y - k)$

Detailed solutions to selected exercises

Exercise 6 For $y^2 - 10x = 0$, find the vertex, the focus, and the equation of the directrix. Also, sketch the parabola.

Solution First, rewrite $y^2 - 10x = 0$ as $y^2 = 10x$. Matching this equation to the form $y^2 = 4px$, we have

$$y^2 = 4(\tfrac{5}{2})x \quad \text{so} \quad p = \tfrac{5}{2}.$$

The focus is on the axis of symmetry (the x-axis) p units to the right of the vertex (the origin). Thus

$$\text{focus: } (\tfrac{5}{2}, 0).$$

The directrix is the vertical line p units to the left of the vertex. Thus

$$\text{directrix: } x = -\tfrac{5}{2}.$$

The parabola is graphed in Figure 8.5:6.

8.5

A. For $y^2 - 4x = 0$, find the vertex, the focus, and the equation of the directrix. Also, sketch the parabola.

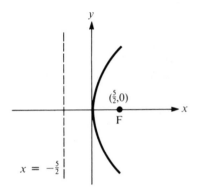

Figure 8.5:6

Now do **A.**

Exercise 14 For $y^2 + 4y + 3x - 8 = 0$, find the vertex, the focus, and the equation of the directrix. Also, sketch the parabola.

Solution First, put the equation in standard form by completing the square.

$$(y^2 + 4y \quad) = -3x + 8$$
$$y^2 + 4y + 4 = -3x + 8 + 4$$
$$(y + 2)^2 = -3(x - 4)$$

Matching this equation to $(y - k)^2 = -4p(x - h)$, we have

$$h = 4, k = -2, -4p = -3 \text{ so } p = \tfrac{3}{4}.$$

Then as shown in the figure for this case,

$$\text{vertex: } (h,k) = (4,-2)$$
$$\text{focus: } (h - p,k) = (\tfrac{13}{4},-2)$$
$$\text{directrix: } x = h + p = \tfrac{19}{4}.$$

The parabola is graphed in Figure 8.5:14.

144

B. For $y^2 + 4y - 4x = 0$, find the vertex, the focus, and the equation of the directrix. Also, sketch the parabola.

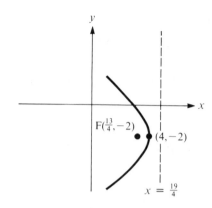

Figure 8.5:14

C. Find an equation of the parabola with focus $(5,-1)$ and directrix $y = 3$.

Now do **B.**

Exercise 22 Find the equation in standard form of the parabola with focus $(-1,-3)$ and directrix $y = 5$.

Solution First, sketch Figure 8.5:22. Note in the figure that the axis of symmetry, $x = -1$, is perpendicular to the directrix and these lines meet at $(-1,5)$. Since the vertex is equidistant from the focus and the directrix, the vertex is at $(-1,1)$ and $p = 4$. Finally, this figure matches the standard form

$$(x - h)^2 = -4p(y - k)$$

so $(x + 1)^2 = -16(y - 1)$ is the standard equation of the parabola.
Now do **C.**

D. Consider the parabolic reflector shown. Choose axes in a convenient position and determine
  a. an equation of the parabola
  b. the location of the focus point

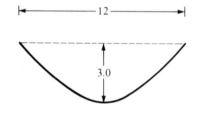

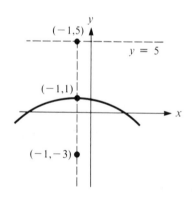

Figure 8.5:22

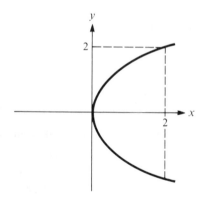

Figure 8.5:26

Exercise 26 See text for question.

Solution First, place the parabola with its vertex at the origin as shown in Figure 8.5:26. The standard form for a parabola in this case is $y^2 = 4px$. Since $(2,2)$ is a point on the parabola, we have

$$2^2 = 4p(2) \quad \text{so} \quad p = \tfrac{1}{2}.$$

Thus, the equation of the parabola is $y^2 = 2x$ and the focus is $\tfrac{1}{2}$ unit to the right of the vertex on the axis of symmetry.
Now do **D.**

8.6 Classifying conic sections

Objectives checklist

Can you:

a. Classify a given equation as defining either a circle, an ellipse, a hyperbola, or a parabola, or a degenerate case of one of these?

Key term

General form (of an equation of a conic section)

Key rules and formulas

- Each of the conic sections can be represented by an equation of the form

$$Ax^2 + Cy^2 + Dx + Ey + F = 0,$$

where A and C are not both zero. This equation is called the general form of an equation of a conic section with axis or axes parallel to the coordinate axes. The distinguishing characteristics for each conic are as follows:

Conditions on A and C	Conic Section	Degenerate Possibilities
$A = C \neq 0$	Circle	A point or no graph at all
$A \neq C$, A and C have the same sign.	Ellipse	A point or no graph at all
A and C have opposite signs.	Hyperbola	Two intersecting straight lines
$A = 0$ or $C = 0$ (but not both)	Parabola	A line, a pair of parallel lines, or no graph at all

Detailed solutions to selected exercises

Exercise 6   Identify the graph of the equation $3y^2 = 4x - x^2$ if the graph is either a circle, an ellipse, a hyperbola, or a parabola.

Solution   Rewrite $3y^2 = 4x - x^2$ as $3y^2 + x^2 - 4x = 0$. Then referring to the general form $Ax^2 + Cy^2 + Dx + Ey + F = 0$, A and C have the same sign with $A \neq C$. Thus, the graph of the equation $3y^2 = 4x - x^2$ is an ellipse. Now do **A**.

Exercise 12   Identify the graph of the equation $x^2 + y^2 - 4y + 4 = 0$. Degenerate cases are possible.

Solution   Referring to the general form of an equation of a conic section, $A = C \neq 0$, so the graph of $x^2 + y^2 - 4y + 4 = 0$ may be a circle. However, to be sure that we are not dealing with a degenerate case, we write the equation in standard form which gives $x^2 + (y - 2)^2 = 0$. Only $x = 0$, $y = 2$ satisfies the equation, so the graph is the point $(0,2)$. Now do **B**.

Sample test questions: Chapter 8

1. Complete the statement.

   a. The standard equation of a circle with center $(-2,4)$ and radius 3 is

   _____ .

   b. The standard equation of a parabola with focus $(0,8)$ and directrix

   $y = -8$ is _____ .

8.6

A. Identify the graph of the equation $3x^2 = 4y + y^2$ as a circle, an ellipse, a hyperbola, or a parabola.

B. Identify the graph of the equation $2x^2 + 2y^2 + 6x + 2y - 3 = 0$. Degenerate cases are possible.

c. The conic section defined as the set of all points in a plane the sum of whose distances from two fixed points is a constant is the _____ .

d. The asymptotes for the hyperbola $x^2 - 4y^2 = 36$ are _____ .

e. The radius of the circle $2x^2 + 4x + 2y^2 + 8y = 36$ is _____ .

f. A hyperbola with center at $(-1,5)$ and one vertex at $(3,5)$ has its other vertex at _____ .

g. The conic section defined by $y^2 = 1 - 2x$ is a _____ .

h. The standard equation of an ellipse with center at the origin, focus $(0,4)$, and major axis of length 14 is _____ .

i. The coordinates of the foci for $25x^2 - 4y^2 = 100$ are _____ .

j. The nondegenerate conic section formed from the intersection of a cone with a plane that is parallel to the base of the cone is the

_____ .

2. Graph each equation.
   a. $4x^2 + y^2 = 1$ b. $x^2 - 9y^2 = 36$ c. $(y + 2)^2 = 4(x + 1)$

3. Find an equation for the graph that is the set of all points in a plane equidistant from $(3,-1)$ and $(-2,4)$.

4. a. An elliptical arch has a height of 25 ft and a span of 60 ft. How high is the arch 10 ft each side of center?

   b. A parabolic arch has a height of 25 ft and a span of 60 ft. How high is the arch 10 ft each side of center?

5. Select the choice that completes the statement or answers the question.

   a. The center of the circle defined by $x^2 + y^2 + 4x - 6y = 0$ is

      (a) $(-2,3)$ (b) $(2,-3)$ (c) $(1,-3)$ (d) $(4,-6)$ (e) $(-4,6)$

   b. The ellipse $x^2 + (y^2/9) = 1$ has one vertex at the point

      (a) $(0,3)$ (b) $(-3,0)$ (c) $(10,0)$ (d) $(0,-10)$ (e) $(9,0)$

   c. The standard equation of the hyperbola with center at the origin and with one vertex at $(0,2)$ and focus at $(0,5)$ is

      (a) $\dfrac{x^2}{4} - \dfrac{y^2}{21} = 1$ (b) $\dfrac{x^2}{21} - \dfrac{y^2}{4} = 1$ (c) $\dfrac{y^2}{4} - \dfrac{x^2}{21} = 1$

      (d) $\dfrac{x^2}{21} + \dfrac{y^2}{4} = 1$ (e) $\dfrac{y^2}{4} + \dfrac{x^2}{21} = 1$

   d. The graph of $6y^2 = 4x - x^2$ is

      (a) a circle   (b) an ellipse   (c) a parabola

      (d) a hyperbola   (e) a point

   e. The standard equation of the circle with center at $(0,2)$ and radius 3 is

      (a) $(x - 2)^2 + y^2 = 3$ (b) $(x + 2)^2 + y^2 = 9$

      (c) $x^2 + (y + 2)^2 = 3$ (d) $x^2 + (y - 2)^2 = 9$

      (e) $x^2 + (y + 2)^2 = 9$

Chapter 9
Systems of Equations and Inequalities

9.1 Systems of linear equations in two variables

Objectives checklist

Can you:

a. Solve a system of linear equations by using the substitution method or the addition-elimination method?

b. Solve applied problems by setting up and solving a system of linear equations?

Key Terms

Linear equation in n variables
Linear system
Inconsistent system of equations
Dependent system of equations

Substitution method
Addition-elimination method
Point of market equilibrium

Key rules and formulas

- The solution set of a system of linear equations is the set of all the ordered pairs that satisfy both equations. Graphically, this corresponds to the set of points where the lines intersect.

- *Substitution method:* To solve a system of equations by this method,

    1. If necessary, solve one of the equations for x or y.
    2. Substitute the resulting expression for the same variable in the other equation and solve.
    3. Use the answer and one of the equations to find the other coordinate in the solution.

- *Addition-elimination method:* To solve a system of equations by this method,

    1. If necessary, form equivalent equations, so that the coefficients of x or y are negatives of each other, by multiplying the equations in the system by nonzero numbers.
    2. Add the equations, thus eliminating one of the variables, and solve for the other variable.
    3. Use this answer and one of the equations to find the other coordinate in the solution.

Detailed solutions to selected exercises

Exercise 8   Find all ordered pairs that satisfy the pair of equations $y = -3x$ and $2x + 3y = -21$.

Solution   By the substitution method, we start with the equation $2x + 3y = -21$ and replace y by $-3x$. Then

$$2x + 3(-3x) = -21 \quad \text{so} \quad -7x = -21 \quad \text{so} \quad x = 3.$$

Using $y = -3x$, when $x = 3$, $y = -3(3) = -9$. Thus, the solution is $(3, -9)$.
Now do **A**.

Chapter 9

9.1

A. Find all ordered pairs that satisfy the pair of equations $y = -5x$ and $3x - 2y = -26$.

148

B. Solve the system of equations
$$2x - 3y = 11$$
$$5x + 8y = 12.$$

C. If one black metal ball and six red metal balls are placed on a scale, they balance a weight of 132 g. A weight of 120 g will balance five black and three red balls. Find the weight of each kind of metal ball.

D. Repeat Exercise 36 assuming machine A will cost \$3,000 per year plus \$3 to package each unit and machine B will cost \$7,000 per year plus \$2 to package each unit. In part (b) assume that packaging can be subcontracted to another firm at a cost of \$4 per unit.

Exercise 18 Solve the system of equations $\begin{cases} 4x + 2y = 2 \\ 6x - 5y = 27. \end{cases}$

Solution We make the coefficients of y negatives of each other by multiplying both sides of $4x + 2y = 2$ by 5 and both sides of $6x - 5y = 27$ by 2. The system then is
$$20x + 10y = 10$$
$$12x - 10y = 54.$$

Adding the equations eliminates y and gives $32x = 64$ so $x = 2$. To find y, substitute 2 for x in $4x + 2y = 2$ giving
$$4(2) + 2y = 2 \quad \text{so} \quad 2y = -6 \quad \text{so} \quad y = -3.$$
Thus, the solution is $(2, -3)$.
Now do **B.**

Exercise 26 See text for question.

Solution If b represents the weight of a black ball and r represents the weight of a red ball, then the problem gives us that
$$4b + r = 100$$
$$2b + 3r = 90.$$

If we multiply both sides of the top equation by -3, we can eliminate r by adding the equations. So,

$$\begin{array}{rcr} -12b - 3r &=& -300 \\ \text{add} \quad\quad 2b + 3r &=& 90 \\ \hline -10b &=& -210 \\ b &=& 21. \end{array}$$

To find r, substitute 21 for b in either of the original equations. Using $4b + r = 100$, we have $4(21) + r = 100$ so $r = 16$. Thus, a black ball weighs 21 g and a red ball weighs 16 g.
Now do **C.**

Exercise 36 See text for question.

Solution
a. The cost equation for x units from Machine A is $C = 2x + 5,000$ while the cost equation for Machine B is $C = x + 8,000$. The cost of the two machines is the same when
$$2x + 5,000 = x + 8,000$$
$$x = 3,000.$$

Thus, the cost is the same for 3,000 units. If more units are needed, Machine B should be used because the cost per unit is only \$1.

b. At \$5 per unit, the cost equation is $C = 5x$. This equation intersects $C = 2x + 5,000$ when
$$5x = 2x + 5,000$$
$$x = 1,667.$$

Thus, purchasing machine A is worthwhile if at least 1,667 units must be produced.
Now do **D.**

9.2 Determinants

Objectives checklist

Can you:

a. Evaluate a determinant with two rows and two columns?

b. Evaluate a determinant with three rows and three columns?

c. Use Cramer's rule to solve systems of two equations in two unknowns or systems of three equations in three unknowns?

Key terms

Determinant

Cramer's rule

Minor of an element

Cofactor of an element

Key rules and formulas

- A determinant is a square array of numbers enclosed by vertical bars, and we define the value of a 2 by 2 determinant as follows:

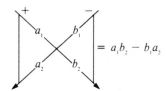 $= a_1 b_2 - b_1 a_2$

- The minor of an element is the determinant formed by deleting the row and column containing the given element.

- The cofactor of the element in the ith row and jth column equals

$$(-1)^{i+j} \cdot \text{(minor of the element)}.$$

- We evaluate 3 by 3 (and more complicated) determinants by following these steps.

   1. Pick any row or column of the determinant.
   2. Multiply each entry in that row or column by its cofactor.
   3. Add the results. This sum is defined to be the value of the determinant.

- *Cramer's rule (two equations with two variables):* The solution to the system

$$\begin{array}{l} a_1 x + b_1 y = c_1 \\ a_2 x + b_2 y = c_2 \end{array} \quad \text{with} \quad a_1 b_2 - b_1 a_2 \neq 0 \quad \text{is}$$

$$x = \frac{D_x}{D} = \frac{\begin{vmatrix} c_1 & b_1 \\ c_2 & b_2 \end{vmatrix}}{\begin{vmatrix} a_1 & b_1 \\ a_2 & b_2 \end{vmatrix}} \qquad y = \frac{D_y}{D} = \frac{\begin{vmatrix} a_1 & c_1 \\ a_2 & c_2 \end{vmatrix}}{\begin{vmatrix} a_1 & b_1 \\ a_2 & b_2 \end{vmatrix}}.$$

9.2

A. Evaluate $\begin{vmatrix} 4 & -21 \\ 3 & -10 \end{vmatrix}$.

B. Evaluate $\begin{vmatrix} 2 & 2 & 4 \\ 3 & 8 & 1 \\ -1 & -1 & -2 \end{vmatrix}$.

C. Solve the system
$$4x - 3y = -26$$
$$-5x + 3y = 7.$$

• *Cramer's rule (three equations with three variables):* The solution to the system

$$\begin{array}{l} a_1x + b_1y + c_1z = d_1 \\ a_2x + b_2y + c_2z = d_2 \\ a_3x + b_3y + c_3z = d_3 \end{array} \quad \text{with } D = \begin{vmatrix} a_1 & b_1 & c_1 \\ a_2 & b_2 & c_2 \\ a_3 & b_3 & c_3 \end{vmatrix} \neq 0$$

is $x = D_x/D$, $y = D_y/D$, and $z = D_z/D$, where

$$D_x = \begin{vmatrix} d_1 & b_1 & c_1 \\ d_2 & b_2 & c_2 \\ d_3 & b_3 & c_3 \end{vmatrix}, D_y = \begin{vmatrix} a_1 & d_1 & c_1 \\ a_2 & d_2 & c_2 \\ a_3 & d_3 & c_3 \end{vmatrix}, D_z = \begin{vmatrix} a_1 & b_1 & d_1 \\ a_2 & b_2 & d_2 \\ a_3 & b_3 & d_3 \end{vmatrix}.$$

Detailed solutions to selected exercises

Exercise 8   See text for question.

Solution   By the given formula,

$$\begin{vmatrix} 5 & -28 \\ 4 & -10 \end{vmatrix} = 5(-10) - (-28)(4) = 62.$$

Now do **A.**

Exercise 18   See text for question.

Solution   The sign pattern for row 1 is $+, -, +$. Using row 1 and following the steps given to evaluate a 3 by 3 determinant, we have

$$\begin{vmatrix} 9 & 10 & -2 \\ 4 & 1 & 8 \\ -1 & -7 & 3 \end{vmatrix} = 9 \begin{vmatrix} 1 & 8 \\ -7 & 3 \end{vmatrix} - 10 \begin{vmatrix} 4 & 8 \\ -1 & 3 \end{vmatrix} + (-2) \begin{vmatrix} 4 & 1 \\ -1 & -7 \end{vmatrix}$$
$$= 9[1(3) - 8(-7)] - 10[4(3) - 8(-1)] - 2[4(-7) - 1(-1)]$$
$$= 9(59) - 10(20) - 2(-27)$$
$$= 385.$$

Now do **B.**

Exercise 24   Solve the system $\begin{cases} -2x - y = -5 \\ 5x + 2y = -17 \end{cases}$ by Cramer's rule.

Solution   By Cramer's rule,

$$x = \frac{\begin{vmatrix} -5 & -1 \\ -17 & 2 \end{vmatrix}}{\begin{vmatrix} -2 & -1 \\ 5 & 2 \end{vmatrix}} = \frac{-5(2) - (-1)(-17)}{-2(2) - (-1)(5)} = \frac{-27}{1} = -27$$

$$y = \frac{\begin{vmatrix} -2 & -5 \\ 5 & -17 \end{vmatrix}}{\begin{vmatrix} -2 & -1 \\ 5 & 2 \end{vmatrix}} = \frac{-2(-17) - (-5)(5)}{1} = \frac{59}{1} = 59.$$

Thus, the solution is $(-27, 59)$.
Now do **C.**

Exercise 34   See text for question.

Solution   The four determinants defined in Cramer's rule are

$$D = \begin{vmatrix} 1 & 1 & -3 \\ -1 & -1 & 5 \\ 2 & 3 & -1 \end{vmatrix} = -2, \quad D_x = \begin{vmatrix} -4 & 1 & -3 \\ 8 & -1 & 5 \\ 3 & 3 & -1 \end{vmatrix} = -2,$$

$$D_y = \begin{vmatrix} 1 & -4 & -3 \\ -1 & 8 & 5 \\ 2 & 3 & -1 \end{vmatrix} = -2, \quad D_z = \begin{vmatrix} 1 & 1 & -4 \\ -1 & -1 & 8 \\ 2 & 3 & 3 \end{vmatrix} = -4.$$

Then $x = \dfrac{D_x}{D} = \dfrac{-2}{-2} = 1$, $y = \dfrac{D_y}{D} = \dfrac{-2}{-2} = 1$, and $z = \dfrac{D_z}{D} = \dfrac{-4}{-2} = 2$.

Now do **D**.

D. Use determinants to solve the
system $-x - y + 2z = 7$
$x + 2y - 2z = -7$
$2x - y + z = -4$.

9.3 Triangular form and matrices

Objectives checklist

Can you:

a. Write a system of equations corresponding to an augmented matrix?

b. Use matrix form to solve a system of equations?

Key terms

Matrix
Coefficient matrix
Augmented matrix

Triangular form
Equivalent systems
Entry (or element) of a matrix

Key rules and formulas

- The following operations may be used to produce equivalent systems of linear equations:

Elementary Operations on Equations
1. Multiply both sides of an equation by a nonzero number.
2. Add a multiple of one equation to another.
3. Interchange two equations.

Elementary Row Operations on Matrices
1. Multiply each entry in a row by a nonzero number.
2. Add a multiple of the entries in one row to another row.
3. Interchange two rows.

- *Gaussian elimination:* A linear system in either equation form or in augmented matrix form may be solved as follows:

Step 1: Reduce the system to triangular form using elementary operations.
Step 2: Solve the last equation. Then, use back substitution to solve the remaining equations.

Note: If you get a row of zeros in the coefficient portion of any matrix, there is no unique solution to the system of equations.

Additional comments

- When using Gaussian elimination, instead of stopping when triangular form is reached, an alternative method is to continue until reaching a form like

$$\begin{bmatrix} 1 & 0 & 0 & | & a \\ 0 & 1 & 0 & | & b \\ 0 & 0 & 1 & | & c \end{bmatrix}.$$

From this, we conclude $x = a, y = b, z = c$.

9.3

A. Solve the system
$$3a - 7b = -2$$
$$5a + 2b = -17$$
and show both the matrix form of the system and the corresponding equations.

Detailed solutions to selected exercises

Exercise 2 Solve the system
$$2a + 5b = 1$$
$$3a - 4b = 13.$$

Solution We use the elementary operations and proceed as follows:

Equation Form   *Matrix Form*

$$2a + 5b = 1$$
$$3a - 4b = 13$$

$$\begin{bmatrix} 2 & 5 & \vdots & 1 \\ 3 & -4 & \vdots & 13 \end{bmatrix}$$

↓ Multiply both sides of the first equation by 3.

↓ Multiply each entry in row 1 by 3.

$$6a + 15b = 3$$
$$3a - 4b = 13$$

$$\begin{bmatrix} 6 & 15 & \vdots & 3 \\ 3 & -4 & \vdots & 13 \end{bmatrix}$$

↓ Multiply both sides of the second equation by -2.

↓ Multiply each entry in row 2 by -2.

$$6a + 15b = 3$$
$$-6a + 8b = -26$$

$$\begin{bmatrix} 6 & 15 & \vdots & 3 \\ -6 & 8 & \vdots & -26 \end{bmatrix}$$

↓ Add the first equation to the second equation.

↓ Add the entries in row 1 to the corresponding entries in row 2.

$$6a + 15b = 3$$
$$23b = -23$$

$$\begin{bmatrix} 6 & 15 & \vdots & 3 \\ 0 & 23 & \vdots & -23 \end{bmatrix}$$

The last row or last equation tells us $23b = -23$, so $b = -1$. Then substituting $b = -1$ into $6a + 15b = 3$ gives $6a - 15 = 3$, so $a = 3$. Thus, the solution is $a = 3$, $b = -1$.

Now do **A.**

Exercise 12 Use matrix form to solve the system
$$x + y + 3z = 1$$
$$2x + 5y + 2z = 0$$
$$3x - 2y - z = 3.$$

Solution The augmented matrix for the system is
$$\begin{bmatrix} 1 & 1 & 3 & \vdots & 1 \\ 2 & 5 & 2 & \vdots & 0 \\ 3 & -2 & -1 & \vdots & -3 \end{bmatrix}.$$

Now use elementary row operations as follows:
Add -2 times each entry in row 1 to the corresponding entry in row 2.
$$\begin{bmatrix} 1 & 1 & 3 & \vdots & 1 \\ 0 & 3 & -4 & \vdots & -2 \\ 3 & -2 & -1 & \vdots & -3 \end{bmatrix}$$

Add -3 times each entry in row 1 to the corresponding entry in row 3.
$$\begin{bmatrix} 1 & 1 & 3 & \vdots & 1 \\ 0 & 3 & -4 & \vdots & -2 \\ 0 & -5 & -10 & \vdots & 0 \end{bmatrix}$$

Multiply each entry in row 2 by 5 and each entry in row 3 by 3.
$$\begin{bmatrix} 1 & 1 & 3 & \vdots & 1 \\ 0 & 15 & -20 & \vdots & -10 \\ 0 & -15 & -30 & \vdots & 0 \end{bmatrix}$$

Add each entry in row 2 to the corresponding entry in row 3.

$$\begin{bmatrix} 1 & 1 & 3 & | & 1 \\ 0 & 15 & -20 & | & -10 \\ 0 & 0 & -50 & | & -10 \end{bmatrix}$$

From the last row, $-50z = -10$, so $z = \frac{1}{5}$. Substitution into the equation corresponding to row 2 gives

$$15y - 20z = -10$$
$$15y - 20(\tfrac{1}{5}) = -10$$
$$y = -\tfrac{6}{15} \text{ or } -\tfrac{2}{5}.$$

Then by substitution of these values for y and z into the equation corresponding to row 1,

$$x + y + 3z = 1$$
$$x - \tfrac{2}{5} + 3(\tfrac{1}{5}) = 1$$
$$x = \tfrac{4}{5}.$$

Thus, the solution is $x = \frac{4}{5}$, $y = -\frac{2}{5}$, $z = \frac{1}{5}$.
Now do **B.**

Exercise 18   See text for question.

Solution   The augmented matrix for the system is

$$\begin{bmatrix} 1 & 1 & 0 & 0 & | & 2 \\ 0 & 1 & 1 & 0 & | & 1 \\ 0 & 0 & 1 & 1 & | & -1 \\ 1 & 0 & 0 & 1 & | & 0 \end{bmatrix}.$$

Now use elementary row operations as follows:
Add -1 times each entry in row 1 to the corresponding entry in row 4.

$$\begin{bmatrix} 1 & 1 & 0 & 0 & | & 2 \\ 0 & 1 & 1 & 0 & | & 1 \\ 0 & 0 & 1 & 1 & | & -1 \\ 0 & -1 & 0 & 1 & | & -2 \end{bmatrix}$$

Add each entry in row 2 to the corresponding entry in row 4.

$$\begin{bmatrix} 1 & 1 & 0 & 0 & | & 2 \\ 0 & 1 & 1 & 0 & | & 1 \\ 0 & 0 & 1 & 1 & | & -1 \\ 0 & 0 & 1 & 1 & | & -1 \end{bmatrix}$$

Add -1 times each entry in row 3 to the corresponding entry in row 4.

$$\begin{bmatrix} 1 & 1 & 0 & 0 & | & 2 \\ 0 & 1 & 1 & 0 & | & 1 \\ 0 & 0 & 1 & 1 & | & -1 \\ 0 & 0 & 0 & 0 & | & 0 \end{bmatrix}$$

Since the entries in the last row are all zeros, the system does not have a unique solution (the system is dependent).
Now do **C.**

Exercise 24   The points $(3,3)$, $(-2,-2)$, and $(1,-1)$ lie on the circle $x^2 + y^2 + Dx + Ey + F = 0$. Find D, E, and F. What are the center and radius of this circle?

B. Use matrix form to solve the system $x + y - 2z = 0$
$2x + 2y - z = 1$
$3x + 2y - 3z = 3.$

C. Use matrix form to solve the system $x + 3z = 14$
$-y + 2w = 7$
$3x + 4w = 26$
$4x - 3y + 2w = 9.$

154

D. Find the equation of the circle $x^2 + y^2 + Dx + Ey + F = 0$ whose graph goes through the points $(-1,2)$, $(3,6)$, and $(7,2)$. Also, find the center and radius of the circle.

Solution   Substituting the values $x = 3$ and $y = 3$, $x = -2$ and $y = -2$, and $x = 1$, $y = -1$, respectively, into the equation of the circle gives the following system of equations.

$$3D + 3E + F = -18$$
$$-2D - 2E + F = -8$$
$$D - E + F = -2$$

The augmented matrix form for the system is

$$\begin{bmatrix} 3 & 3 & 1 & | & -18 \\ -2 & -2 & 1 & | & -8 \\ 1 & -1 & 1 & | & -2 \end{bmatrix}.$$

Interchange the entries in row 3 with the corresponding entries in row 1.

$$\begin{bmatrix} 1 & -1 & 1 & | & -2 \\ -2 & -2 & 1 & | & -8 \\ 3 & 3 & 1 & | & -18 \end{bmatrix}$$

Add 2 times each entry in row 1 to the corresponding entries in row 2.

$$\begin{bmatrix} 1 & -1 & 1 & | & -2 \\ 0 & -4 & 3 & | & -12 \\ 3 & 3 & 1 & | & -18 \end{bmatrix}$$

Add -3 times each entry in row 1 to the corresponding entries in row 3.

$$\begin{bmatrix} 1 & -1 & 1 & | & -2 \\ 0 & -4 & 3 & | & -12 \\ 0 & 6 & -2 & | & -12 \end{bmatrix}$$

Multiply each entry in row 2 by 3 and each entry in row 3 by 2.

$$\begin{bmatrix} 1 & -1 & 1 & | & -2 \\ 0 & -12 & 9 & | & -36 \\ 0 & 12 & -4 & | & -24 \end{bmatrix}$$

Add each entry in row 2 to the corresponding entry in row 3.

$$\begin{bmatrix} 1 & -1 & 1 & | & -2 \\ 0 & -12 & 9 & | & -36 \\ 0 & 0 & 5 & | & -60 \end{bmatrix}$$

From the last row, $F = -12$. Then substituting this value into the equation corresponding to row 2 gives

$$-12E + 9F = -36$$
$$-12E + 9(-12) = -36$$
$$E = -6.$$

Substituting these values for E and F into the equation corresponding to row 1 gives

$$D - E + F = -2$$
$$D - (-6) - 12 = -2$$
$$D = 4.$$

Since $D = 4$, $E = -6$, and $F = -12$, the equation of the circle is $x^2 + y^2 + 4x - 6y - 12 = 0$. Completing the square gives the standard equation $(x + 2)^2 + (y - 3)^2 = 25$, from which we conclude that the center of the circle is $(-2,3)$ and the radius is 5.
Now do **D.**

9.4 Solving systems by matrix algebra

Objectives checklist

Can you:

a. Determine the dimension of a matrix?

b. Determine the values that make two matrices equal?

c. Determine if the product of two matrices is defined and, if so, predict the dimension of the product?

d. Add, subtract, and multiply matrices and multiply a matrix by a scalar?

e. Find the inverse of a matrix, if it exists?

f. Write the system of linear equations represented in matrix form by $AX = B$?

g. Write a system of linear equations in matrix form $AX = B$ and use A^{-1} to solve the system?

Key terms

Matrix
Square matrix
Scalar
$n \times n$ identity matrix
Equal matrices

Entry (or element) of a matrix
Zero matrix
Principal diagonal
Multiplicative inverse of A

Key rules and formulas

- *Matrix operations:*

    1. Matrix addition (or subtraction)—To add (or subtract) two matrices, add (or subtract) elements in corresponding positions. Only matrices of the same dimension may be added (or subtracted).
    2. Scalar multiplication—To multiply a matrix by a real number (scalar), multiply each element in the matrix by that number.
    3. Matrix multiplication—If A is an $m \times n$ matrix and B is an $n \times p$ matrix, then the product AB is an $m \times p$ matrix in which the ith row, jth column element of AB is found by multiplying each element in the ith row of A by the corresponding element in the jth column of B and adding the results. Note that AB is defined only when the number of columns in A matches the number of rows in B, and that AB has as many rows as A and as many columns as B.

- *Properties of matrices:* If A, B, and C are matrices and c and k are real numbers, then

    1. $(A + B) + C = A + (B + C)$
    2. $A + B = B + A$
    3. $A + 0 = 0 + A = A$
    4. $A + (-A) = (-A) + A = 0$
    5. $c(kA) = (ck)A$
    6. $c(A + B) = cA + cB.$

These properties carry over to matrices from the real numbers. Note, however, that matrix multiplication is *not* commutative (in general, $AB \neq BA$).

9.4

A. What is the dimension of the following matrix?

$$\begin{bmatrix} -6 \\ 0 \\ 8 \\ 2 \end{bmatrix}$$

B. Find the values of the variables, if

$$\begin{bmatrix} 3a + 1 & -4b \\ -c - 2 & 3d - 1 \\ 2e + 3 & -2f - 3 \end{bmatrix} = \begin{bmatrix} -2 & 8 \\ -6 & 0 \\ 5 & -7 \end{bmatrix}.$$

C. Perform the indicated operations.

$$2\begin{bmatrix} -5 \\ 0 \\ 1 \end{bmatrix} + \frac{1}{4}\begin{bmatrix} 8 \\ 7 \\ -4 \end{bmatrix}$$

D. If A is a 4×2 matrix and B is a 2×2 matrix, find the dimensions of the product AB and the product BA if they are defined.

- The multiplicative inverse of matrix A (symbolized A^{-1}) exists if and only if the determinant of A is not zero. To find A^{-1}:

Step 1: Write the augmented matrix $[A \mid I]$, where I is the identity matrix with the same dimension as A.

Step 2: Use elementary row operations to replace matrix $[A \mid I]$ with a matrix of the form $[I \mid B]$.

Step 3: Then A^{-1} is matrix B.

If it's not possible to obtain $[I \mid B]$, then A has no inverse.

- The solution to the linear system $AX = B$ is given by $X = A^{-1}B$, provided A^{-1} exists.

Detailed solutions to selected exercises

Exercise 4   What is the dimension of the matrix $[1 \quad 3 \quad 5]$?

Solution   The matrix has 1 row and 3 columns, so it is a 1×3 matrix. Now do **A.**

Exercise 8   Find the values of the variables if

$$\begin{bmatrix} 2 + 2a & -1 + 2b & 3 + 2c \\ 2d & 1 + 2e & -2 + 2f \end{bmatrix} = \begin{bmatrix} 4 & 0 & 1 \\ -2 & 7 & 2 \end{bmatrix}.$$

Solution   If two matrices are equal, the corresponding entries are equal. Thus,

$2 + 2a = 4$, so $a = 1$; $-1 + 2b = 0$, so $b = \frac{1}{2}$; $3 + 2c = 1$, so $c = -1$;
$2d = -2$, so $d = -1$; $1 + 2e = 7$, so $e = 3$; $-2 + 2f = 2$, so $f = 2$.

Now do **B.**

Exercise 14   Perform the indicated operations.

$$-\tfrac{1}{2}[2 \quad -4 \quad 6] + \tfrac{1}{3}[-9 \quad 3 \quad 6]$$

Solution   First, we do the scalar multiplications, multiplying each entry in the first matrix by $-\frac{1}{2}$ and each element in the second matrix by $\frac{1}{3}$.

$$-\tfrac{1}{2}[2 \quad -4 \quad 6] + \tfrac{1}{3}[-9 \quad 3 \quad 6]$$
$$= [-1 \quad 2 \quad -3] + [-3 \quad 1 \quad 2]$$

Now adding corresponding entries gives

$$[-1 \quad 2 \quad -3] + [-3 \quad 1 \quad 2] = [-4 \quad 3 \quad -1].$$

Now do **C.**

Exercise 20   If A is a 3×3 matrix and B is a 1×3 matrix, find the dimensions of the product AB and the product BA if they are defined.

Solution A has 3 columns and B has 1 row. Since these numbers are not equal, the product AB is not defined. B has 3 columns and A has 3 rows. Since these numbers are equal, the product BA is defined. The dimension of BA is 1×3, because B has 1 row and A has 3 columns. Now do **D.**

Exercise 24   Determine the product

$$\begin{bmatrix} 1 & 3 \\ -1 & 0 \end{bmatrix} \begin{bmatrix} 2 \\ 3 \end{bmatrix}.$$

Solution The product of a 2 × 2 matrix with a 2 × 1 matrix is a 2 × 1 matrix. The product is found as follows:

$$\begin{bmatrix} 1 & 3 \\ -1 & 0 \end{bmatrix} \begin{bmatrix} 2 \\ 3 \end{bmatrix} = \begin{bmatrix} (1)(2) + (3)(3) \\ (-1)(2) + (0)(3) \end{bmatrix} = \begin{bmatrix} 11 \\ -2 \end{bmatrix}.$$

Now do **E**.

E. Determine the product.

$$\begin{bmatrix} 3 & 4 & -2 \\ -1 & 2 & 5 \end{bmatrix} \begin{bmatrix} 1 \\ 2 \\ 0 \end{bmatrix}$$

Exercise 34 If $D = \begin{bmatrix} 1 & 3 & 6 \\ 2 & 5 & 7 \end{bmatrix}$ and $A = \begin{bmatrix} 2 & -1 & 3 \\ 0 & 1 & -2 \end{bmatrix}$, find $3D - 2A$.

Solution By direct substitution,

$$3D - 2A = 3\begin{bmatrix} 1 & 3 & 6 \\ 2 & 5 & 7 \end{bmatrix} - 2\begin{bmatrix} 2 & -1 & 3 \\ 0 & 1 & -2 \end{bmatrix}$$

then

$$3D - 2A = \begin{bmatrix} 3 & 9 & 18 \\ 6 & 15 & 21 \end{bmatrix} - \begin{bmatrix} 4 & -2 & 6 \\ 0 & 2 & -4 \end{bmatrix}$$

$$= \begin{bmatrix} -1 & 11 & 12 \\ 6 & 13 & 25 \end{bmatrix}.$$

F. If $A = \begin{bmatrix} 2 & 3 & 1 \\ 0 & -4 & 5 \end{bmatrix}$ and $B = \begin{bmatrix} -5 & 2 & 4 \\ 3 & 0 & -1 \end{bmatrix}$, find $2A - 3B$.

Now do **F**.

Exercise 40 Write the system of linear equations represented by $AX = B$ if

$$A = \begin{bmatrix} 3 & -1 & 5 \\ 1 & 0 & -2 \\ -1 & 1 & -1 \end{bmatrix}, X = \begin{bmatrix} x \\ y \\ z \end{bmatrix}, \text{ and } B = \begin{bmatrix} 1 \\ -1 \\ 1 \end{bmatrix}.$$

Solution Since A is a 3 × 3 matrix and X is a 3 × 1 matrix, the product AX is a 3 × 1 matrix.

$$AX = \begin{bmatrix} 3 & -1 & 5 \\ 1 & 0 & -2 \\ -1 & 1 & -1 \end{bmatrix} \begin{bmatrix} x \\ y \\ z \end{bmatrix} = \begin{bmatrix} 3x + (-1y) + 5z \\ 1x + 0y + (-2z) \\ -1x + 1y + (-1z) \end{bmatrix}$$

$$= \begin{bmatrix} 3x - y + 5z \\ x - 2z \\ -x + y - z \end{bmatrix}$$

G. Write the system of linear equations represented by $AX = B$ if

$$A = \begin{bmatrix} 2 & -1 & 3 \\ 0 & 5 & -9 \\ -4 & 3 & -1 \end{bmatrix},$$

$$X = \begin{bmatrix} x \\ y \\ z \end{bmatrix}, \text{ and }$$

$$B = \begin{bmatrix} 0 \\ 2 \\ -1 \end{bmatrix}.$$

Since $AX = B$, we know

$$\begin{bmatrix} 3x - y + 5z \\ x - 2z \\ -x + y - z \end{bmatrix} = \begin{bmatrix} 1 \\ -1 \\ 1 \end{bmatrix}.$$

Then from the definition of equal matrices,

$$\begin{aligned} 3x - y + 5z &= 1 \\ x \qquad - 2z &= -1 \\ -x + y - z &= 1. \end{aligned}$$

Now do **G**.

Exercise 48 Find the inverse of the matrix $\begin{bmatrix} 2 & 3 & 0 \\ 3 & 0 & 2 \\ 0 & 4 & 1 \end{bmatrix}$ if it exists.

H. Find the inverse of the matrix

$$\begin{bmatrix} -1 & 2 & -3 \\ 2 & 1 & 0 \\ 4 & -2 & 5 \end{bmatrix}, \text{ if it exists.}$$

Solution To find A^{-1}, form the augmented matrix $[A \mid I]$ and convert it to the form $[I \mid B]$ as follows:

$$\begin{bmatrix} 2 & 3 & 0 & \vdots & 1 & 0 & 0 \\ 3 & 0 & 2 & \vdots & 0 & 1 & 0 \\ 0 & 4 & 1 & \vdots & 0 & 0 & 1 \end{bmatrix}.$$

Multiply each entry in row 1 by $\frac{1}{2}$.

$$\begin{bmatrix} 1 & \frac{3}{2} & 0 & \vdots & \frac{1}{2} & 0 & 0 \\ 3 & 0 & 2 & \vdots & 0 & 1 & 0 \\ 0 & 4 & 1 & \vdots & 0 & 0 & 1 \end{bmatrix}$$

Add -3 times each entry in row 1 to the corresponding entry in row 2.

$$\begin{bmatrix} 1 & \frac{3}{2} & 0 & \vdots & \frac{1}{2} & 0 & 0 \\ 0 & -\frac{9}{2} & 2 & \vdots & -\frac{3}{2} & 1 & 0 \\ 0 & 4 & 1 & \vdots & 0 & 0 & 1 \end{bmatrix}$$

Multiply each entry in row 2 by $-\frac{2}{9}$.

$$\begin{bmatrix} 1 & \frac{3}{2} & 0 & \vdots & \frac{1}{2} & 0 & 0 \\ 0 & 1 & -\frac{4}{9} & \vdots & \frac{1}{3} & -\frac{2}{9} & 0 \\ 0 & 4 & 1 & \vdots & 0 & 0 & 1 \end{bmatrix}$$

Add $-\frac{3}{2}$ times each entry in row 2 to the corresponding entry in row 1.

$$\begin{bmatrix} 1 & 0 & \frac{2}{3} & \vdots & 0 & \frac{1}{3} & 0 \\ 0 & 1 & -\frac{4}{9} & \vdots & \frac{1}{3} & -\frac{2}{9} & 0 \\ 0 & 4 & 1 & \vdots & 0 & 0 & 1 \end{bmatrix}$$

Add -4 times each entry in row 2 to the corresponding entry in row 3.

$$\begin{bmatrix} 1 & 0 & \frac{2}{3} & \vdots & 0 & \frac{1}{3} & 0 \\ 0 & 1 & -\frac{4}{9} & \vdots & \frac{1}{3} & -\frac{2}{9} & 0 \\ 0 & 0 & \frac{25}{9} & \vdots & -\frac{4}{3} & \frac{8}{9} & 1 \end{bmatrix}$$

Multiply each entry in row 3 by $\frac{9}{25}$.

$$\begin{bmatrix} 1 & 0 & \frac{2}{3} & \vdots & 0 & \frac{1}{3} & 0 \\ 0 & 1 & -\frac{4}{9} & \vdots & \frac{1}{3} & -\frac{2}{9} & 0 \\ 0 & 0 & 1 & \vdots & -\frac{12}{25} & \frac{8}{25} & \frac{9}{25} \end{bmatrix}$$

Add $-\frac{2}{3}$ times each entry in row 3 to the corresponding entry in row 1 and $\frac{4}{9}$ times each entry in row 3 to the corresponding entry in row 2.

$$\begin{bmatrix} 1 & 0 & 0 & \vdots & \frac{8}{25} & \frac{3}{25} & -\frac{6}{25} \\ 0 & 1 & 0 & \vdots & \frac{3}{25} & -\frac{2}{25} & \frac{4}{25} \\ 0 & 0 & 1 & \vdots & -\frac{12}{25} & \frac{8}{25} & \frac{9}{25} \end{bmatrix}$$

Thus,

$$A^{-1} = \begin{bmatrix} \frac{8}{25} & \frac{3}{25} & -\frac{6}{25} \\ \frac{3}{25} & -\frac{2}{25} & \frac{4}{25} \\ -\frac{12}{25} & \frac{8}{25} & \frac{9}{25} \end{bmatrix} \text{ or } \frac{1}{25} \begin{bmatrix} 8 & 3 & -6 \\ 3 & -2 & 4 \\ -12 & 8 & 9 \end{bmatrix}.$$

Now do **H.**

Exercise 56. Write the given system in the form $AX = B$ and use A^{-1} to solve the system. (*Note:* The inverse for this problem is found in Exercise 47.)

$$x - y + z = 1$$
$$-x + y = -1$$
$$ -y + z = 2$$

Solution The given system can be represented by $AX = B$, where

$$A = \begin{bmatrix} 1 & -1 & 1 \\ -1 & 1 & 0 \\ 0 & -1 & 1 \end{bmatrix}, X = \begin{bmatrix} x \\ y \\ z \end{bmatrix}, \text{ and } B = \begin{bmatrix} 1 \\ -1 \\ 2 \end{bmatrix}.$$

If A^{-1} exists, the solution to $AX = B$ is $X = A^{-1}B$. We know from Exercise 47 that

$$A^{-1} = \begin{bmatrix} 1 & 0 & -1 \\ 1 & 1 & -1 \\ 1 & 1 & 0 \end{bmatrix}.$$

So

$$X = \begin{bmatrix} x \\ y \\ z \end{bmatrix} = \begin{bmatrix} 1 & 0 & -1 \\ 1 & 1 & -1 \\ 1 & 1 & 0 \end{bmatrix} \begin{bmatrix} 1 \\ -1 \\ 2 \end{bmatrix}$$

$$= \begin{bmatrix} (1)(1) + (0)(-1) + (-1)(2) \\ (1)(1) + (1)(-1) + (-1)(2) \\ (1)(1) + (1)(-1) + (0)(-2) \end{bmatrix} = \begin{bmatrix} -1 \\ -2 \\ 0 \end{bmatrix}.$$

Thus, $x = -1$, $y = -2$, and $z = 0$.
Now do **I**.

I. Write the given system in the form $AX = B$.

$$x + 2y + 3z = 0$$
$$2x + 3y + 4z = 2$$
$$3x + 4y + 6z = 3$$

Use the information that

$$A^{-1} = \begin{bmatrix} -2 & 0 & 1 \\ 0 & 3 & -1 \\ 1 & -2 & 1 \end{bmatrix}$$

to solve the system.

9.5 Partial fractions

Objectives checklist

Can you:

a. Complete the partial fraction decomposition of $P(x)/Q(x)$ by determining the constants in a given equation that make the equation an identity?

b. Determine the partial fraction decomposition of $P(x)/Q(x)$ in the cases in which linear and/or quadratic factors are repeated in the denominator, as well as when no factors are repeated?

Key terms

Partial fraction   Partial fraction decomposition
Irreducible quadratic factor

Key rules or formulas

• To determine the partial fraction decomposition of $P(x)/Q(x)$, first divide out any common factors of $P(x)$ and $Q(x)$. Then do the following:

1. If the degree of $P(x)$ is not less than the degree of $Q(x)$, use long division to rewrite the fraction as

$$\frac{P(x)}{Q(x)} = \text{quotient} + \frac{\text{remainder}}{Q(x)}.$$

9.5

A. Determine the constants A, B, C, and D so that the given equation is an identity.

$$\frac{3x^3 - 10x^2 + 24x - 18}{x^2(x^2 - 3x + 9)}$$
$$= \frac{A}{x} + \frac{B}{x^2} + \frac{Cx + D}{x^2 - 3x + 9}$$

2. Now work with the remainder term.
   a. Factor completely the denominator $Q(x)$ into linear factors and/or irreducible quadratic factors.
   b. Each factor of the denominator will produce one or more terms in the partial fraction decomposition according to the following rules:
   Linear factors: Each linear factor of $Q(x)$ of the form $(ax + b)^n$ produces in the decomposition a sum of n terms of the form

$$\frac{A_1}{ax + b} + \frac{A_2}{(ax + b)^2} + \cdots + \frac{A_n}{(ax + b)^n}.$$

   Note that when a linear factor appears just once, it simply produces a single term of the form $\frac{A}{ax + b}$.

   Quadratic factors: Each irreducible quadratic factor of $Q(x)$ of the form $(ax^2 + bx + c)^n$ produces a sum of n terms of the form

$$\frac{A_1x + B_1}{ax^2 + bx + c} + \frac{A_2x + B_2}{(ax^2 + bx + c)^2} + \cdots + \frac{A_nx + B_n}{(ax^2 + bx + c)^n}.$$

   Note that when an irreducible quadratic factor appears just once, it simply produces a single term of the form $\frac{Ax + B}{ax^2 + bx + c}$.

   c. Determine the values of the constants in the numerators of the partial fractions by multiplying both sides of the equation by the least common denominator, equating coefficients of like powers of x, and solving the resulting system of equations.

Additional comments

- Instead of using subscript notation, we usually use the letters A, B, C, and so on to denote the constants in the numerators.

Detailed solutions to selected exercises

Exercise 10 Determine the constants A, B, C, and D so that the equation is an identity.

$$\frac{2x^3 - 5x^2 + 4x - 3}{x^2(x^2 + 1)} = \frac{A}{x} + \frac{B}{x^2} + \frac{Cx + D}{x^2 + 1}$$

Solution Multiplying both sides of the equation by $x^2(x^2 + 1)$ and expanding gives

$$\begin{aligned}
2x^3 - 5x^2 + 4x - 3 &= A(x)(x^2 + 1) + B(x^2 + 1) + (Cx + D)(x^2) \\
&= Ax^3 + Ax + Bx^2 + B + Cx^3 + Dx^2 \\
&= (A + C)x^3 + (B + D)x^2 + Ax + B.
\end{aligned}$$

Equating coefficients of like powers of x gives the system

$$\begin{aligned}
A + C &= 2 \\
B + D &= -5 \\
A &= 4 \\
B &= -3.
\end{aligned}$$

Solving the system, we get $A = 4$, $B = -3$, $C = -2$, and $D = -2$. Now do **A.**

Exercise 16 Find the partial fraction decomposition of $\dfrac{x^2}{x^2 - 2x + 1}$.

Solution Since the degree of the numerator is not less than the degree of the denominator, we use long division, which yields

$$\frac{x^2}{x^2 - 2x + 1} = 1 + \frac{2x - 1}{x^2 - 2x + 1}.$$

Factoring the denominator of the remainder term gives

$$\frac{2x - 1}{x^2 - 2x + 1} = \frac{2x - 1}{(x - 1)^2}.$$

Since $(x - 1)^2$ is a repeated linear factor, we write

$$\frac{2x - 1}{(x - 1)^2} = \frac{A}{x - 1} + \frac{B}{(x - 1)^2}.$$

Then clearing denominators gives

$$\begin{aligned} 2x - 1 &= A(x - 1) + B \\ &= Ax - A + B. \end{aligned}$$

Equating coefficients gives $A = 2$ and $-A + B = -1$, which implies $B = 1$. Thus,

$$\frac{x^2}{x^2 - 2x + 1} = 1 + \frac{2}{x - 1} + \frac{1}{(x - 1)^2}.$$

Now do **B**.

Exercise 22 Find the partial fraction decomposition of $\dfrac{1}{x^3 - 1}$.

Solution Since the degree of the numerator is less than the degree of the denominator, we begin by factoring the denominator, which yields

$$\frac{1}{x^3 - 1} = \frac{1}{(x - 1)(x^2 + x + 1)}.$$

Since $(x - 1)$ is a nonrepeated linear factor and $(x^2 + x + 1)$ is a nonrepeated quadratic factor, we represent the fraction as follows:

$$\frac{1}{(x - 1)(x^2 + x + 1)} = \frac{A}{x - 1} + \frac{Bx + C}{x^2 + x + 1}.$$

Then

$$\begin{aligned} 1 &= A(x^2 + x + 1) + (Bx + C)(x - 1) \\ &= Ax^2 + Ax + A + Bx^2 + Cx - Bx - C \\ &= (A + B)x^2 + (A + C - B)x + A - C \end{aligned}$$

Equating coefficients gives the system

$$\begin{aligned} A + B &= 0 \\ A + C - B &= 0 \\ A - C &= 1. \end{aligned}$$

The solution of this system is $A = \frac{1}{3}$, $B = -\frac{1}{3}$, $C = -\frac{2}{3}$. Thus,

$$\frac{1}{x^3 - 1} = \frac{\frac{1}{3}}{x - 1} + \frac{-\frac{1}{3}x - \frac{2}{3}}{x^2 + x + 1}.$$

Now do **C**.

B. Find the partial fraction decomposition of $\dfrac{4x^2 + 10x}{4x^2 + 4x + 1}$.

C. Determine the partial fraction decomposition of

$$\frac{5}{x^3 + 2x^2 + 2x + 1}.$$

9.6 Nonlinear systems of equations

Objectives checklist

Can you:

a. Solve nonlinear systems of equations by the substitution method or the addition-elimination method?

b. Find the intersection points of two curves represented by a nonlinear system of equations?

c. Solve applied problems by setting up and solving a nonlinear system of equations?

Key terms

Substitution method   Addition-elimination method

Key rules and formulas

• The solution set of a nonlinear system of equations is the set of all the ordered pairs that satisfy both equations. Graphically, this corresponds to the set of points where the curves intersect.

• *Substitution method:* To solve a system of equations by this method,

  1. If necessary, solve one of the equations for x or y, or for the lowest appearing power of the variable.
  2. Substitute the resulting expression for the same variable in the other equation and solve.
  3. Use the answer and one of the equations to find the other coordinate in the solution.

• *Addition-elimination method:* To solve a system of equations by this method,

  1. If necessary, form equivalent equations, so that the coefficients of like powers of x or y are negatives of each other, by multiplying the equations in the system by nonzero numbers.
  2. Add the equations, thus eliminating one of the variables, and solve for the other variable.
  3. Use this answer and one of the equations to find the other coordinate in the solution.

Additional comments

• The substitution and addition-elimination methods for solving nonlinear systems of equations are basically the same as for solving systems of linear equations.

Detailed solutions to selected exercises

Exercise 16   Find all the ordered pairs that satisfy the equations

$$xy = 4$$
$$x^2 + y^2 = 8.$$

Solution   We solve the first equation for x and rewrite the system as

$$x = \frac{4}{y}$$
$$x^2 + y^2 = 8.$$

Then by the substitution method, we replace x by $4/y$ in the second equation, giving $(4/y)^2 + y^2 = 8$. Now solve for y as follows:

$$\frac{16}{y^2} + y^2 = 8, \text{ so } 16 + y^4 = 8y^2, \text{ so } y^4 - 8y^2 + 16 = 0.$$

Factoring gives $(y^2 - 4)^2 = 0$, so $y^2 = 4$, so $y = \pm 2$. Since $xy = 4$, $y = 2$ implies $x = 2$ and $y = -2$ implies $x = -2$. Thus, $(2,2)$ and $(-2,-2)$ satisfy both equations.
Now do **A**.

Exercise 26 Find all ordered pairs that satisfy the equations

$$\frac{x^2}{4} + y^2 = 1$$
$$\frac{x^2}{4} - \frac{y^2}{2} = 1.$$

Solution Subtracting the second equation from the first eliminates the x variable, yielding $y^2 + (y^2/2) = 0$. Solving for y, we get $3y^2/2 = 0$, so $y = 0$. Then substituting $y = 0$ into $(x^2/4) + y^2 = 1$ and solving for x, we have $x^2/4 = 1$, so $x = \pm 2$. Thus, $(2,0)$ and $(-2,0)$ satisfy both equations.
Now do **B**.

Exercise 32 Find all intersection points of the semicircle $y = -\sqrt{36 - x^2}$ and the line $y = -x$.

Solution We need to solve the system

$$y = -\sqrt{36 - x^2}$$
$$y = -x.$$

By the substitution method, we replace y by $-x$ in the first equation, giving $-x = -\sqrt{36 - x^2}$. Then $x^2 = 36 - x^2$, so $x^2 = 18$, giving $x = \pm 3\sqrt{2}$. Since $y = -x$, $x = 3\sqrt{2}$ implies $y = -3\sqrt{2}$ and $x = -3\sqrt{2}$ implies $y = 3\sqrt{2}$. However, we reject the solution $(-3\sqrt{2}, 3\sqrt{2})$, since checking these values in $y = -\sqrt{36 - x^2}$ shows that the solution is extraneous. Thus, the given semicircle and line intersect in the single point $(3\sqrt{2}, -3\sqrt{2})$.
Now do **C**.

Exercise 38 See text for question.

Solution Let r_1 be the radius of the first circle and r_2 be the radius of the second circle. Then

$$r_1 + r_2 = 9$$
$$\pi(r_1)^2 + \pi(r_2)^2 = 48.5\pi.$$

From the first equation, $r_1 = 9 - r_2$. Replacing r_1 in the second equation, we have

$$\pi(9 - r_2)^2 + \pi(r_2)^2 = 48.5\pi$$
$$(9 - r_2)^2 + (r_2)^2 = 48.5$$
$$81 - 18r_2 + (r_2)^2 + (r_2)^2 = 48.5$$
$$2(r_2)^2 - 18r_2 + 32.5 = 0.$$

By the quadratic formula,

$$r_2 = \frac{18 \pm \sqrt{324 - 260}}{4} \text{ so } r_2 = 2.5 \text{ or } 6.5.$$

9.6

A. Find all ordered pairs that satisfy the equations

$$xy = -9$$
$$x^2 - y^2 = -80.$$

B. Find all ordered pairs that satisfy the equations

$$x^2 - \frac{y^2}{9} = 3$$
$$\frac{x^2}{4} + \frac{y^2}{9} = 2.$$

C. Find all intersection points of the line $y = x - 1$ and the semicircle $y = -\sqrt{13 - x^2}$.

D. If the radius of one circle is twice the radius of the second circle and the combined areas of the two circles is 151.25π square inches, find the radius of each circle.

E. Alan and Bill together can paint a room in 4 hours, Alan and Cora together can paint the room in 2 hours, and Bill and Cora together can paint it in 2 hours and 24 minutes. How long does it take each person alone to paint the room? How long does it take for the room to be painted if all three people paint?

Since $r_1 + r_2 = 9$, if $r_2 = 2.5$ then $r_1 = 6.5$, and vice versa. Thus, the radii of the circles are 2.5 inches and 6.5 inches.
Now do **D.**

Exercise 42   See text for question.

Solution   Let $x =$ the number of hours in which A can fill the tank working alone

$y =$ the number of hours in which B can fill the tank working alone

$z =$ the number of hours in which C can fill the tank working alone.

In 1 hour the portions of the tank A, B, and C can each fill are $1/x$, $1/y$, and $1/z$, respectively. From the given information on the time needed to complete the whole job, we get the system of equations

$$
\begin{aligned}
7.5(1/x) + 7.5(1/y) \qquad\quad &= 1 \\
6(1/x) \qquad\quad + 6(1/z) &= 1 \\
10(1/y) + 10(1/z) &= 1.
\end{aligned}
$$

Letting $a = 1/z$, $b = 1/y$, and $c = 1/z$, we get the equivalent system

$$
\begin{aligned}
7.5a + 7.5b \qquad\quad &= 1 \\
6a \qquad\quad + 6c &= 1 \\
10b + 10c &= 1.
\end{aligned}
$$

Solving the system by any of the methods discussed in previous sections (for example, Gaussian elimination) yields $a = \frac{1}{10}$, $b = \frac{1}{30}$, and $c = \frac{1}{15}$, which implies $x = 10$, $y = 30$, and $z = 15$. Thus, it takes tank A 10 hours, tank B 30 hours, and tank C 15 hours to fill the tank.

If all three pumps are used, we let n represent the number of hours it takes to fill the tanks. Then

$$ n(\tfrac{1}{10}) + n(\tfrac{1}{30}) + n(\tfrac{1}{15}) = 1, \text{ so } 3n + n + 2n = 30, \text{ so } n = 5. $$

Thus, it takes 5 hours to fill the tank using all three pumps.
Now do **E.**

9.7 Systems of linear inequalities and linear programming

Objectives checklist

Can you:

a. Graph the solution set of an inequality?

b. Graph the solution set of a system of inequalities?

c. Find the maximum and/or minimum values of an objective function subject to given constraints?

d. Find an inequality that expresses a given relationship?

e. Solve applied linear programming problems by first setting up an objective function and a system of constraints?

Key terms

Linear programming

Vertices (or corners) of a region

Objective function

Feasible solution

Simplex method

Half-plane

Test point

Constraints

Convex region

Key rules and formulas

- *Graphing inequalities in two variables:* Replace the inequality symbol with an equal sign and graph the resulting equality to get a line. The solution set of the inequality consists of one of the two resulting half-planes and includes the line if the inequality symbol is $\leq$ or $\geq$. A test point may be used to determine which half-plane to shade in. A solid line is used when the line is included in the solution set, otherwise a dashed line is used.

- The solution set of a system of inequalities is the intersection of the solution sets of all the individual inequalities in the system. To find the solution graphically, graph the solution set of each inequality on the same coordinate system. Then the solution set of the system is the intersection of these half-planes.

- *Corner point theorem:* If a linear function $F = ax + by$ assumes a maximum or minimum value subject to a system of linear inequalities, then it does so at the corners or vertices of that system of constraints. Furthermore, as long as the set of feasible solutions is a closed convex polygon, we are guaranteed that a maximum or minimum exists.

- To find the maximum and/or minimum value of an objective function subject to given constraints, do the following:

  1. Graph the solution set of the system of inequalities formed by the constraints and specify the coordinates of any corner in the graph.
  2. If the set of feasible solutions is a closed convex polygon, a maximum or minimum value exists at one of the corners. Substitute the coordinates of the vertices into the objective function to determine the correct corner.

Detailed solutions to selected exercises

Exercise 8   Graph the inequality $x + 2y \leq 5$. Use shading to indicate the graph.

Solution   We first graph the line $x + 2y = 5$. Then substituting the coordinates of the test point $(0,0)$ into the inequality gives the true statement $0 \leq 5$. Therefore we shade the half-plane which includes the origin to get the graph shown in Figure 9.7:8.

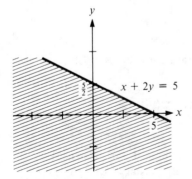

Figure 9.7:8

Now do **A.**

9.7

A. Graph the inequality $2x - y \leq 7$. Use shading to indicate the graph.

B. Graph the solution set of the system of inequalities

$$x + y \leq 7$$
$$x - 2y \leq -8.$$

Specify the coordinates of any corner in the graph.

C. Find the maximum and minimum values of $F = 3x + 4y$ subject to the constraints

$$x + y \leq 7$$
$$x - 2y \leq -8$$
$$x \geq 0$$
$$y \geq 0.$$

(See Margin Exercise B.)

Exercise 14   See text for question.

Solution   We first graph each of the lines $x + y = 1$, $x - y = 1$, and $x + 2y = 2$ on the same set of axes. Then by using (0,0) as a test point for each inequality, we determine that $x + 2y \leq 2$ is satisfied by the points on or below the line $x + 2y = 2$, whereas $x + y \geq 1$ and $x - y \leq 1$ are satisfied by the points on or above the lines $x + y = 1$ and $x - y = 1$, respectively. The intersection of these half-planes is shown in Figure 9.7:14. By solving each of the systems

$$\begin{array}{ccc} x + y = 1 & \text{and} & x + y = 1 & \text{and} & x - y = 1 \\ x - y = 1 & & x + 2y = 2 & & x + 2y = 2 \end{array}$$

using substitution or addition-elimination, we determine that the corner points are (1,0), (0,1) and $(\frac{4}{3}, \frac{1}{3})$.

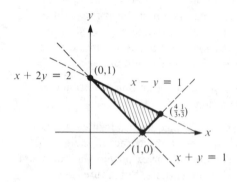

Figure 9.7:14

Now do **B.**

Exercise 18   Find the maximum and minimum values of $F = 6x - 12y$ subject to the constraints given in Exercise 14.

Solution   We substitute the coordinates of the vertices obtained in Exercise 14 into $F = 6x - 12y$ as follows:

Vertex	$F = 6x - 12y$
(1,0)	$6(1) - 12(0) = 6$
(0,1)	$6(0) - 12(1) = -12$
$(\frac{4}{3}, \frac{1}{3})$	$6(\frac{4}{3}) - 12(\frac{1}{3}) = 4$

Thus, the maximum value of F is 6 and the minimum value is -12. Now do **C.**

Exercise 24   Maximize $P = 15x + 8y + 9$ subject to $x - y \leq 7$
$$x - y \geq 1$$
$$8 \leq x \leq 13.$$

Solution   Graphically solving the system to determine the set of feasible solutions, we obtain the closed convex polygon with vertices (8,1), (8,7), (13,6), and (13,12) shown in Figure 9.7:24.

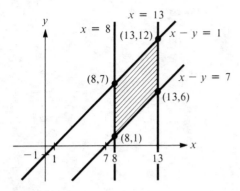

Figure 9.7:24

The value of P at each of these vertices is as follows:

Vertex	$P = 15x + 8y + 9$
(8,1)	137
(8,7)	185
(13,6)	252
(13,12)	300

Thus, the maximum value of P is 300.
Now do **D**.

Exercise 28   See text for question.

Solution   Since Sam should consume at least 2 mg of thiamine daily, we have $1.0x + 0.7y \geq 2$. Similarly, since he should consume at least 25 mg of niacin daily, we have $10.0x + 12.0y \geq 25$. Also, x and y cannot be negative so $x \geq 0$ and $y \geq 0$.
Now do **E**.

Exercise 32   See text for question.

Solution   We want to minimize the cost subject to the constraints imposed by the minimum daily requirements and the fact that x and y cannot be negative. The cost function is

$$C = 0.15x + 0.13y$$

and the constraints are

$$1.0x + 0.7y \geq 2$$
$$10.0x + 12.0y \geq 25$$
$$x \geq 0$$
$$y \geq 0.$$

We solve the system of inequalities to determine the set of feasible solutions shown in Figure 9.7:32. Note that the corner point (1.3,1) is obtained by solving the system of equations $1.0x + 0.7y = 2$ and $10.0x + 12.0y = 25$.

D. Maximize $P = 10x + 4y + 2$
subject to

$$x + y \leq 10$$
$$x + y \geq 0$$
$$1 \leq x \leq 6.$$

E. A nutritionist wishes to combine two foods, A and B, into a diet with at least the minimum requirement of vitamins (160 units) and of minerals (100 units). The following is known about the two foods.

Food	Vitamins per Ounce	Minerals per Ounce
A	4	1
B	2	2

If x represents the number of ounces used of food A and y represents the number of ounces of food B, find two inequalities that express what must be done to produce a diet which satisfies the minimum requirement.

F. Use the information given in Margin Exercise E and determine the combination of foods the nutritionist should use to satisfy the minimum requirements of vitamins and minerals at least cost if food A costs 20 cents per ounce and food B costs 15 cents per ounce.

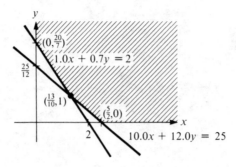

Figure 9.7:32

The value of C at each of the corner points is as follows:

Vertex	$C = 0.15x + 0.13y$
$(0,\frac{20}{7})$	0.371
$(\frac{13}{10},1)$	0.325
$(\frac{5}{2},0)$	0.375

Thus, the minimum cost occurs when Sam eats $\frac{13}{10}$ or 1.3 ounces of cereal A and 1 ounce of cereal B.
Now do **F.**

Sample test questions: Chapter 9

1. Find the value of the determinants.

   a. $\begin{vmatrix} -2 & -4 \\ 3 & 1 \end{vmatrix}$ b. $\begin{vmatrix} 3 & -2 & 0 \\ 0 & 1 & 4 \\ -1 & 5 & 2 \end{vmatrix}$

2. If

$$A = \begin{bmatrix} 2 & -1 \\ 3 & 2 \end{bmatrix}, B = \begin{bmatrix} -1 & 5 \\ 4 & -3 \end{bmatrix}, C = \begin{bmatrix} -1 & 0 \\ 0 & 4 \\ 2 & -2 \end{bmatrix},$$

$$D = \begin{bmatrix} 1 & 3 & 4 \\ -1 & 0 & -6 \end{bmatrix},$$

   then if possible find

   a. $A + C$ b. $B - A$ c. $2D$ d. CD

3. Find the inverse of the matrix $\begin{bmatrix} -2 & 3 \\ 1 & 4 \end{bmatrix}$.

4. Solve each of the following systems of equations by the method indicated.

   a. $2x + 3y = 5$
      $4x + 7y = 11$ (addition-elimination)

   b. $4x + 9y = 8$
      $8x + 6z = -1$ (Cramer's rule)
      $6y + 6z = -1$

5. Solve each system of equations. Use any appropriate method.

a. $8x^2 - y^2 = 71$

$4x^2 + y^2 = 37$

b. $\dfrac{3}{x} - \dfrac{2}{y} + 1 = -1$

$\dfrac{1}{x} + \dfrac{6}{y} + 3 = 4$

c. $-x + 2y + z = 0$

$x + y + z = 1$

$3x - y - z = 7$

6. Solve each applied problem.

a. The sum of two numbers is 2 and the difference of their squares is 12. Find the two numbers.

b. Admission to a concert was $5 for students and $7 for nonstudents. There were 525 paid admissions and total receipts were $3,053. How many students attended the concert?

7. Graph the solution set of this system of inequalities.

$$x + y \leq 7$$
$$x - 2y \leq -8$$

8. Determine the constants A, B, and C so that this equation is an identity.

$$\frac{3x^2 - 10x + 4}{x(x - 2)^2} = \frac{A}{x} + \frac{B}{x - 2} + \frac{C}{(x - 2)^2}$$

9. Find the partial fraction decomposition of the given expression.

a. $\dfrac{2x - 7}{2x^2 + x - 3}$

b. $\dfrac{2x^2 - x - 1}{x^3 + 3x^2 + 3x}$

10. Maximize $P = -x_1 + x_2$ subject to $x_1 + x_2 \leq 10$

$$5x_1 + 16x_2 \leq 80$$
$$x_1, x_2 \geq 0.$$

11. A cattle rancher can buy food mix A at 20¢ per pound and food mix B at 40¢ per pound. Each pound of mix A contains 3,000 units of nutrient N_1 and 1,000 units of N_2; each pound of mix B contains 4,000 units of nutrient N_1 and 4,000 units of nutrient N_2. If the minimum daily requirements for the cattle collectively are 36,000 units of nutrient N_1 and 20,000 units of nutrient N_2, how many pounds of each food mix should be used each day to minimize daily food costs while meeting (or exceeding) the minimum daily nutrient requirements? What is the minimum daily cost?

Chapter 10
Discrete Algebra and Probability

10.1 Sequences

Objectives checklist

Can you:

a. Find any term in a sequence when given a formula for the nth term of the sequence?

b. Determine if a given sequence is an arithmetic progression, geometric progression, or neither?

c. Find the formula for the general term a_n in a given arithmetic progression?

d. Find the formula for the general term a_n in a given geometric progression?

e. State the general rule for the Fibonacci sequence and write a specified number of terms in the sequence?

Key terms

Sequence
Finite sequence
Infinite sequence
Terms (of the sequence)
Fibonacci sequence

Arithmetic progression
Common difference
Geometric progression
Common ratio

Key rules and formulas

- An arithmetic progression is a sequence of numbers in which each term after the first is found by adding a constant to the preceding term. This constant is called the common difference and is symbolized by d. The formula for the nth term in an arithmetic progression is

$$a_n = a_1 + (n - 1)d.$$

- A geometric progression is a sequence of numbers in which each number after the first is found by multiplying the preceding term by a constant. This constant is called the common ratio and is symbolized by r. The formula for the nth term in a geometric progression is

$$a_n = a_1 r^{n-1}.$$

- The Fibonacci sequence is a specific sequence in which the first two terms are both 1, and thereafter each term is the sum of the two previous terms. In symbols, we write

$$a_1 = 1, a_2 = 1, a_n = a_{n-1} + a_{n-2} \text{ for } n \geq 3.$$

Detailed solutions to selected exercises

Exercise 6 Write the first four terms of the sequence given by
$a_n = \dfrac{(-1)^n}{n}$; also find a_{10}.

Solution Substituting $n = 1,2,3,4$ in the rule for a_n gives

$$a_1 = \frac{(-1)^1}{1} = -1 \qquad a_2 = \frac{(-1)^2}{2} = \frac{1}{2}$$

$$a_3 = \frac{(-1)^3}{3} = -\frac{1}{3} \qquad a_4 = \frac{(-1)^4}{4} = \frac{1}{4}.$$

For the 10th term, a_{10}, we have $a_{10} = \frac{(-1)^{10}}{10} = \frac{1}{10}$.

Now do **A**.

Exercise 18 See text for question.

Solution Each term after the first is found by adding 0.01 to the preceding term. Therefore, the sequence is an arithmetic progression in which the common difference is 0.01. The next two terms in the sequence are then 1.06 and 1.07.

Now do **B**.

Exercise 30 Find the formula for the general term in the arithmetic progression $\frac{13}{4}, \frac{9}{2}, \frac{23}{4}, \ldots$; also find a_{20}.

Solution The 1st term is $a_1 = \frac{13}{4}$ and the common difference is $d = \frac{9}{2} - \frac{13}{4} = \frac{5}{4}$. Now substitute these numbers as follows:

$$\begin{aligned} a_n &= a_1 + (n-1)d \\ &= \tfrac{13}{4} + (n-1)\tfrac{5}{4} \\ &= \tfrac{5}{4}n + 2. \end{aligned}$$

To find the 20th term, a_{20}, replace n by 20.

$$a_{20} = \tfrac{5}{4}(20) + 2 = 27$$

Now do **C**.

Exercise 38 Find the formula for the general term in the geometric progression $1, 0.1, 0.01, \ldots$; also find a_{10}.

Solution The 1st term is $a_1 = 1$ and the common ratio is $r = 0.1/1 = 0.1$. Now substitute these numbers as follows:

$$\begin{aligned} a_n &= a_1 r^{n-1} \\ &= 1(0.1)^{n-1}. \end{aligned}$$

For the 10th term, a_{10}, we have

$$a_{10} = (0.1)^{10-1} = (0.1)^9 = 0.000000001.$$

Now do **D**.

Exercise 50 If, in an arithmetic progression, $a_{15} = \frac{22}{3}$ and $a_{30} = \frac{91}{6}$, find a_{40}.

Solution First, find the formula for the general term by setting up a system of equations using the given corresponding values of the variables.

$$\begin{array}{ccc} \frac{22}{3} = a_1 + (15-1)d & \quad 22 = 3a_1 + 42d & \quad -44 = -6a_1 - 84d \\ \frac{91}{6} = a_1 + (30-1)d & \text{so} \quad 91 = 6a_1 + 174d & \text{so} \quad 91 = 6a_1 + 174d \end{array}$$

Then by the addition-elimination method, $d = \frac{47}{90}$. To find a_1, replace d by $\frac{47}{90}$ in one of the equations and solve for a_1.

$$\tfrac{22}{3} = a_1 + 14 \cdot \tfrac{47}{90} \quad \text{so} \quad a_1 = \tfrac{22}{3} - 14 \cdot \tfrac{47}{90} = \tfrac{2}{90} = \tfrac{1}{45}$$

Chapter 10

10.1

A. Write the first 4 terms of the sequence given by

$a_n = \dfrac{2^n}{n^2}$. Also find a_{10}.

B. Is the sequence 11.01, 12.02, 13.03, 14.04, 15.05, . . . arithmetic, geometric, or neither? Determine the common difference or common ratio, if either exists, and write the next two terms.

C. Find the formula for the general term in the arithmetic progression $-\frac{8}{9}, -\frac{1}{3}, \frac{2}{9}, \ldots$; also find a_{10}.

D. Find the formula for the general term in the geometric progression, $1, \frac{1}{4}, \frac{1}{16}, \ldots$; also find a_8.

E. If, in an arithmetic progression, $a_{13} = \frac{67}{12}$ and $a_{25} = \frac{133}{12}$, find a_{49}.

The general formula is $a_n = \frac{1}{45} + (n-1)\frac{47}{90}$ and for a_{40} we have

$$a_{40} = \frac{1}{45} + (40-1)\frac{47}{90} = \frac{1,835}{90} = \frac{367}{18}.$$

Now do **E.**

10.2 Series

Objectives checklist

Can you:

a. Find the sum of an indicated number of terms in a given arithmetic or geometric progression?

b. Write a series given in sigma notation in its expanded form, and determine the sum?

c. Write a series given in expanded form using sigma notation?

d. Solve applied problems involving series?

Key terms

Series
Arithmetic series
Geometric series

Sigma notation
Annuity
Ordinary annuity

Key rules and formulas

- The sum of the first n terms of an arithmetic series is given by

$$S_n = \frac{n}{2}(a_1 + a_n) \quad \text{or} \quad S_n = \frac{n}{2}[2a_1 + (n-1)d].$$

- The sum of the first n terms of a geometric series is given by

$$S_n = \frac{a_1 - a_1 r^n}{1 - r} \quad \text{or} \quad S_n = \frac{a_1 - a_n r}{1 - r}.$$

- *Sigma notation:* The Greek letter Σ (read "sigma") is used to simplify the notation involved with series. By this convention,

$$S_n = \sum_{i=1}^{n} a_i = a_1 + a_2 + \cdots + a_n$$

so that $\displaystyle\sum_{i=1}^{n} a_i$ means to ADD the terms that result from replacing i by 1, then 2, . . . , then n.

Detailed solutions to selected exercises

Exercise 10 Find the sum of the first 10 terms of the sequence $\frac{13}{4}, \frac{9}{2}, \frac{23}{4}, \ldots$.

Solution This sequence is an arithmetic progression in which $a_1 = \frac{13}{4}$ and the common difference is $d = \frac{9}{2} - \frac{13}{4} = \frac{5}{4}$. Then

$$S_n = \frac{n}{2}[2a_1 + (n-1)d] = \frac{10}{2}\left[2 \cdot \frac{13}{4} + (10-1) \cdot \frac{5}{4}\right]$$
$$= 5\left[\frac{26}{4} + \frac{45}{4}\right] = \frac{355}{4}.$$

Now do **A.**

10.2

A. Find the sum of the first 12 terms of the sequence $\frac{5}{6}, 2, \frac{19}{6}, \ldots$.

Exercise 18   Write the series $\sum\limits_{j=2}^{8} 2^j$ in expanded form and determine the sum.

Solution   Replace j by 2, then 3, . . . , then 8, and add the terms together.

$$\sum_{j=2}^{8} 2^j = 2^2 + 2^3 + 2^4 + 2^5 + 2^6 + 2^7 + 2^8$$

This series is a geometric series with $a_1 = 2^2$, $a_n = 2^8$, and $r = 2$. Thus,

$$S_n = \frac{a_1 - a_n r}{1 - r} = \frac{2^2 - 2^8(2)}{1 - 2} = \frac{4 - 512}{-1} = 508.$$

Now do **B**.

Exercise 26   Write the series $\frac{2}{3} + \frac{3}{5} + \frac{4}{7} + \frac{5}{9} + \frac{6}{11}$ in sigma notation.

Solution   Treat the numerator and the denominator separately. As i ranges from 1 to 5 inclusive, the sequence 2, 3, 4, 5, 6 has general term $i + 1$, while the sequence 3, 5, 7, 9, 11 has general term $2i + 1$. Therefore, the given series may be expressed as

$$\sum_{i=1}^{5} \frac{i+1}{2i+1} \cdot \left(\textit{Note: } \text{A given series can be expressed in sigma notation in more} \right.$$

than one way. Two other solutions are $\sum\limits_{i=2}^{6} \dfrac{i}{2i-1}$ and $\left.\sum\limits_{i=0}^{4} \dfrac{i+2}{2i+3} \cdot \right)$

Now do **C**.

Exercise 34   You accept a position at a salary of $10,000 for the first year with an increase of $1,000 per year each year thereafter. How many years will you have to work for your total earnings to equal $231,000?

Solution   The series is an arithmetic series with $a_1 = 10,000$ and a common difference $d = 1,000$. Then

$$S_n = \frac{n}{2}[2a_1 + (n-1)d]$$

$$231,000 = \frac{n}{2}[2(10,000) + (n-1)1,000]$$

$$462,000 = 20,000n + 1,000n^2 - 1,000n$$

$$0 = n^2 + 19n - 462$$

$$0 = (n + 33)(n - 14).$$

Setting each factor equal to zero gives 14 years as the solution.
Now do **D**.

Exercise 36   A certain ball always rebounds $\frac{1}{2}$ as far as it falls. If the ball is thrown 10 ft into the air, how far up and down has it traveled when it hits the ground for the sixth time?

Solution   The first time the ball hits the ground it has traveled 20 ft. The distance for the second trip is 10 ft, the third trip 5 ft, and so on. The total distance for six trips is a geometric series with $n = 6$, $a_1 = 20$, and $r = \frac{1}{2} = 0.5$. Then

$$S_n = \frac{a_1 - a_1 r^n}{1 - r} = \frac{20 - 20(0.5)^6}{1 - 0.5} = \frac{19.6875}{0.5} = 39.375.$$

Thus, the total distance for six bounces is 39.375 ft.
Now do **E**.

B. Write the series $\sum\limits_{j=3}^{7} 10^j$ in expanded form and determine the sum.

C. Write the series
$\frac{3}{4} + \frac{5}{7} + \frac{7}{10} + \frac{9}{13} + \frac{11}{16} + \frac{13}{19}$
in sigma notation.

D. You accept a position at a salary of $18,000 for the first year with an increase of $500 per year each year thereafter. How many years will you have to work for your total earnings to equal $400,500?

E. A certain ball always rebounds $\frac{2}{5}$ as far as it falls. If the ball is thrown 8 ft into the air, how far up and down has it traveled when it hits the ground for the fifth time?

10.3

A. Find the sum of the infinite geometric series
$$\frac{1}{4} - \frac{1}{10} + \frac{1}{25} - \frac{2}{125} + \ldots.$$

B. Express $0.\overline{15}$ as the ratio of two integers.

C. For what values of x does
$2 + 2x + 2x^2 + 2x^3 + \ldots$
converge to a sum? What is this sum?

10.3 Infinite geometric series

Objectives checklist

Can you:

a. Find the sum of an infinite geometric series?

b. Use an infinite geometric series to express a repeating decimal as the ratio of two integers?

Key rules and formulas

- An infinite geometric series with $|r| < 1$ converges to the value or sum

$$S = \frac{a_1}{1 - r}.$$

Additional comments

- To say S is the sum of an infinite series does not mean we add up all the numbers in the series and get S; it means we can get S_n as close to S as we wish merely by taking a sufficiently large value for n.

Detailed solutions to selected exercises

Exercise 8   Find the sum of the infinite geometric series

$$\frac{1}{2} - \frac{1}{3} + \frac{2}{9} - \frac{4}{27} + \cdots.$$

Solution   Here $a_1 = \frac{1}{2}$ and $r = \dfrac{-\frac{1}{3}}{\frac{1}{2}} = -\frac{2}{3}$. Since r is between -1 and 1 we can assign a sum to the series. Then

$$S = \frac{a_1}{1 - r} = \frac{\frac{1}{2}}{1 - (-\frac{2}{3})} = \frac{\frac{1}{2}}{\frac{5}{3}} = \frac{3}{10}.$$

Now do **A**.

Exercise 14   Express $0.0\overline{7}$ as the ratio of two integers.

Solution   The repeating decimal $0.0\overline{7}$ can be written as

$$0.07 + 0.007 + 0.0007 + \cdots.$$

This series is an infinite geometric series with $a_1 = 0.07$ and $r = 0.1$ so

$$S = \frac{a_1}{1 - r} = \frac{0.07}{1 - (0.1)} = \frac{0.07}{0.9} = \frac{7}{90}.$$

Thus, $0.0\overline{7}$ is equivalent to $\frac{7}{90}$.
Now do **B**.

Exercise 22   For what values of x does the series

$$1 + x + x^2 + x^3 + \cdots$$

converge to a sum? What is this sum?

Solution   The series $1 + x + x^2 + x^3 + \cdots$ is an infinite geometric series with $a_1 = 1$ and $r = x$, which converges to a sum provided $|r| < 1$. Thus, for any x in the interval $(-1, 1)$, the sum of this series is

$$S = \frac{a_1}{1 - r} = \frac{1}{1 - x}.$$

Now do **C**.

10.4 Mathematical induction

Objectives checklist

Can you:

a. Use mathematical induction to prove that a statement is true for all positive integers?

b. Use mathematical induction to prove that a statement is true for all integers greater than or equal to a particular integer?

Key terms

Induction   Induction assumption
Mathematical induction

Key rules and formulas

- *Principle of mathematical induction:* If a given statement S_n concerning positive integer n is true for $n = 1$, and if its truth for $n = k$ implies its truth for $n = k + 1$, then S_n is true for every positive integer n.

- To prove by mathematical induction that a statement (or formula) is true for all positive integers, do the following:

  1. By direct substitution, show that the statement is true for $n = 1$.
  2. Show that if the statement is true for any positive integer k, then it is also true for the next highest integer $k + 1$.

- To prove by mathematical induction that a statement is true for all integers greater than or equal to some particular integer q, do the following:

  1. By direct substitution, show that the statement is true for $n = q$.
  2. Show that if the statement is true for any positive integer $k \geq q$, then it is also true for the next highest integer $k + 1$.

Additional comments

- In general, two different methods can be used in Step 2 of a proof to establish the "$k + 1$" statement from the assumption of the "k" statement:

  Method 1: Start with one side of the "$k + 1$" statement and derive the other side with the aid of the "k" statement.

  Method 2: Start with the "k" statement and perform a valid arithmetic or algebraic operation on it to establish the "$k + 1$" statement.

Detailed solutions to selected exercises

Exercise 10   Prove that the formula

$$1^2 + 2^2 + 3^2 + \cdots + n^2 = \frac{n(n + 1)(2n + 1)}{6}$$

is true for all positive integer values of n by mathematical induction.

Solution   Since we are proving the formula true for all positive integers, we proceed as follows:

Step 1 Show the formula true for $n = 1$ by direct substitution. If $n = 1$, $1^2 + 2^2 + 3^2 + \cdots + n^2$ is just 1^2. Then

$$1^2 = \frac{1(1 + 1)[2(1) + 1]}{6} \quad \text{since both sides equal 1.}$$

10.4

A. Prove by mathematical induction that

$$(1 \cdot 2) + (2 \cdot 3) + (3 \cdot 4) + \cdots + n(n + 1)$$
$$= \frac{n(n + 1)(n + 2)}{3}.$$

B. Prove by mathematical induction that $n^3 + 14n + 3$ is divisible by 3 for all positive integers n.

Step 2 Assume the formula true for $n = k$, and then show that it is true for $n = k + 1$. Setting $n = k$ gives our induction assumption,

$$1^2 + 2^2 + 3^2 + \cdots + k^2 = \frac{k(k + 1)(2k + 1)}{6}.$$

Now show that

$$1^2 + 2^2 + \cdots + k^2 + (k + 1)^2 = \frac{(k + 1)[(k + 1) + 1][2(k + 1) + 1]}{6}.$$

Using method 1, derive the right-hand side of this equation from the left-hand side by first substituting $1^2 + 2^2 + 3^2 + \cdots + k^2 = \dfrac{k(k + 1)(2k + 1)}{6}$ (induction assumption). Then

$$1^2 + 2^2 + 3^2 + \cdots + k^2 + (k + 1)^2$$
$$= \frac{k(k + 1)(2k + 1)}{6} + (k + 1)^2$$
$$= (k + 1)\left[\frac{k(2k + 1)}{6} + (k + 1)\right]$$
$$= (k + 1)\left[\frac{2k^2 + k}{6} + \frac{6k + 6}{6}\right]$$
$$= (k + 1)\left[\frac{2k^2 + 7k + 6}{6}\right]$$
$$= \frac{(k + 1)(k + 2)(2k + 3)}{6}$$
$$= \frac{(k + 1)[(k + 1) + 1][2(k + 1) + 1]}{6}.$$

Now do **A.**

Exercise 18 Prove by mathematical induction that $n^2 + n$ is divisible by 2 for all positive integers n.

Solution Since we are proving a statement true for all positive integers, we proceed as follows:

Step 1 Show that the statement is true for $n = 1$. Direct substitution yields $1^2 + 1 = 2$, which is divisible by 2.

Step 2 Assume the statement true for $n = k$ and then show that it is true for $n = k + 1$. Setting $n = k$ gives the induction assumption that $k^2 + k$ is divisible by 2. Now show that $(k + 1)^2 + (k + 1)$ is divisible by 2. To use the induction assumption, we rewrite this expression in terms of $k^2 + k$. So,

$$(k + 1)^2 + (k + 1) = (k^2 + 2k + 1) + (k + 1)$$
$$= [(k^2 + k) + (k + 1)] + (k + 1)$$
$$= (k^2 + k) + 2(k + 1).$$

By the induction assumption, $k^2 + k$ is divisible by 2. Also, $2(k + 1)$ is obviously divisible by 2. It follows that $(k + 1)^2 + (k + 1)$ is divisible by 2. Now do **B.**

10.5 Binomial theorem

Objectives checklist

Can you:

a. Use the binomial theorem to expand $(a + b)^n$ for any positive integer n?

b. Find the rth term in the expansion of $(a + b)^n$?

c. Express the entries in Pascal's triangle using the notation $\binom{n}{r}$?

d. Evaluate binomial coefficients in the form $\binom{n}{r}$ for given values of n and r?

e. Use the binomial coefficient formula and/or binomial expansions to prove certain properties of binomial coefficients?

Key terms

Pascal's triangle   Factorial notation

Key rules and formulas

- The following array of numbers is known as Pascal's triangle.

Row 0:   1

Row 1:   1   1

Row 2:   1   2   1

Row 3:   1   3   3   1

Row 4:   1   4   6   4   1

Row 5:   1   5   10   10   5   1

This array specifies the constant coefficients in the expansion of $(a + b)^n$ for $n = 0, 1, 2, \ldots$. Except for the 1s, each entry in Pascal's triangle is the sum of the two numbers on either side of it in the preceding row as diagrammed in the chart.

- For any positive integer n, the symbol $n!$ (read "n factorial") means the product $n \cdot (n - 1) \cdot (n - 2) \cdots 3 \cdot 2 \cdot 1$.

- *Binomial theorem:* For any positive integer n,

$$(a + b)^n = a^n + \frac{n}{1!}a^{n-1}b + \frac{n(n-1)}{2!}a^{n-2}b^2 + \cdots$$
$$+ \frac{n(n-1)(n-2)\cdots(n-r+1)}{r!}a^{n-r}b^r + \cdots + b^n.$$

10.5

A. Expand $(3c + d)^4$.

- *Binomial coefficient:* Let r and n be nonnegative integers with $r \le n$. Then the symbol $\binom{n}{r}$ is defined by

$$\binom{n}{r} = \frac{n!}{r!(n-r)!}.$$

Each of the numbers $\binom{n}{r}$ is called a binomial coefficient.

- *Binomial theorem (alternate version):* For any positive integer n,

$$(a + b)^n = \sum_{r=0}^{n} \binom{n}{r} a^{n-r} b^r$$

$$= \binom{n}{0} a^n + \binom{n}{1} a^{n-1} b + \binom{n}{2} a^{n-2} b^2 + \cdots$$

$$+ \binom{n}{n-1} ab^{n-1} + \binom{n}{n} b^n.$$

- The rth term of the expansion of $(a + b)^n$ is

$$\binom{n}{r-1} a^{n-(r-1)} b^{r-1}.$$

Additional comments

- The binomial coefficients $\binom{n}{0}, \binom{n}{1}, \binom{n}{2}, \ldots, \binom{n}{n}$ are precisely the numbers in the nth row of Pascal's triangle.

- By using a scientific calculator with a factorial key $\boxed{x!}$, it is easy to evaluate binomial coefficients. Consult the text for an example.

- There are many patterns in the expansion of $(a + b)^n$ for both the constant coefficients and the powers of a and b in the expanded form. Consult the text for some of these useful patterns.

Detailed solutions to selected exercises

Exercise 12 Expand $(4c + 3d)^3$.

Solution In our statement of the binomial theorem, we substitute $4c$ for a and $3d$ for b. By Pascal's triangle (row 3), we determine the coefficients of the four terms in the expansion to be 1, 3, 3, and 1, respectively. Thus,

$$(4c + 3d)^3 = (4c)^3 + 3(4c)^2(3d) + 3(4c)(3d)^2 + (3d)^3$$
$$= 64c^3 + 144c^2d + 108cd^2 + 27d^3.$$

The factorial formula for the coefficients gives the same result as follows:

$$(4c + 3d)^3 = (4c)^3 + \frac{3}{1!}(4c)^2(3d) + \frac{3 \cdot 2}{2!}(4c)(3d)^2 + (3d)^3$$
$$= (4c)^3 + 3(4c)^2(3d) + 3(4c)(3d)^2 + (3d)^3$$
$$= 64c^3 + 144c^2d + 108cd^2 + 27d^3.$$

We can also obtain this result by applying the binomial theorem with $4c = a$, $3d = b$, and $n = 3$ as follows:

$$(4c + 3d)^3 = \binom{3}{0}(4c)^3 + \binom{3}{1}(4c)^2(3d) + \binom{3}{2}(4c)(3d)^2 + \binom{3}{3}(3d)^3$$
$$= 1(4c)^3 + 3(4c)^2(3d) + 3(4c)(3d)^2 + 1(3d)^3$$
$$= 64c^3 + 144c^2d + 108cd^2 + 27d^3.$$

Now do **A.**

Exercise 16   Write the first four terms in the expansion of $(x + 3)^{17}$.

Solution   Here $n = 17$, $a = x$, and $b = 3$. Then by the binomial theorem, the first four terms are

$$x^{17} + \frac{17}{1!}x^{16}(3)^1 + \frac{17 \cdot 16}{2!}x^{15}(3)^2 + \frac{17 \cdot 16 \cdot 15}{3!}x^{14}(3)^3$$

which simplifies to

$$x^{17} + 51x^{16} + 1{,}224x^{15} + 18{,}360x^{14}.$$

Now do **B**.

Exercise 20   Express the entries in row 2 of Pascal's triangle using the notation $\binom{n}{r}$.

Solution   The entries in row 2 of Pascal's triangle are 1, 2, and 1. Using the notation $\binom{n}{r}$ with $n = 2$ (for row 2), we have $1 = \binom{2}{0}$, $2 = \binom{2}{1}$, and $1 = \binom{2}{2}$.

Now do **C**.

Exercise 24   Evaluate the binomial coefficient $\binom{10}{3}$.

Solution   By applying the formula $\binom{n}{r} = \frac{n!}{r!(n-r)!}$ with $n = 10$ and $r = 3$, we have $\binom{10}{3} = \frac{10!}{3!7!} = 120$.

Now do **D**.

Exercise 32   Find the 5th term in the expansion of $(x - y)^7$.

Solution   We use the formula for the rth term of the binomial expansion of $(a + b)^n$, with $n = 7$, $r = 5$, $a = x$, and $b = -y$. Thus, the 5th term is

$$\binom{7}{4}x^{7-(5-1)}(-y)^{5-1} = 35x^3y^4.$$

Now do **E**.

Exercise 48   Show that $\binom{n}{r} = \binom{n}{n-r}$.

Solution   By definition,

$$\binom{n}{r} = \frac{n!}{r!(n-r)!}$$

Replacing r by $n - r$ in this formula gives

$$\binom{n}{n-r} = \frac{n!}{(n-r)!(n-(n-r))!} = \frac{n!}{(n-r)!r!}$$

Thus,

$$\binom{n}{r} = \binom{n}{n-r}.$$

B. Write the first 4 terms in the expansion of $(x + 2)^{10}$.

C. Express the entries in row 3 of Pascal's triangle using the notation $\binom{n}{r}$.

D. Evaluate the binomial coefficient $\binom{9}{6}$.

E. Find the 4th term in the expansion of $(x - 2y)^5$.

10.6 Counting techniques

Objectives checklist

Can you:

a. Evaluate expressions of the form $_nC_r$ and $_nP_r$?

b. Determine the number of ways two (or more) events taken together can occur?

c. Determine the number of distinct permutations of n objects in the case when all the objects are different and in the case when we cannot distinguish between certain members in the set of objects?

d. Determine the number of permutations of n objects taken r at a time?

e. Determine the number of combinations of n objects taken r at a time?

f. Solve applied problems by choosing the appropriate counting techniques and applying the associated formulas?

Key terms

Tree diagram   Combination
Permutation

Key rules and formulas

- *Fundamental counting principle:* If event A can occur a ways, and following this, event B can occur in b ways, then the two events taken together can occur in $a \cdot b$ ways. (*Note:* A tree diagram can be used to list the possibilities.) This principle extends to three or more events.

- A permutation of a set of objects or symbols is an arrangement of these objects, without repetition, in which order is important. The number of permutations of n objects using all of them is

$$n! = n(n - 1)(n - 2) \cdots 3 \cdot 2 \cdot 1.$$

- When we cannot distinguish between certain members in our set of objects, not all of the permutations are distinct. The number of distinguishable permutations of n objects when n_1 are of one kind, n_2 are of a second kind, . . . , and n_k are of a kth kind, where $n_1 + n_2 + \cdots + n_k = n$, is

$$\frac{n!}{n_1!n_2! \cdots n_k!}.$$

- If $_nP_r$ represents the number of permutations of n objects taken r at a time, then

$$_nP_r = \underbrace{n(n - 1)(n - 2) \cdots (n - r + 1)}_{r \text{ factors}} \qquad r \leq n$$

Equivalently,

$$_nP_r = \frac{n!}{(n - r)!}.$$

- A combination of a set of objects or symbols is a selection without repetition in which the order of selection does not matter. If $_nC_r$ represents the number of combinations of n objects taken r at a time, then

$$_nC_r = \frac{_nP_r}{r!} = \frac{n!}{(n - r)!r!}.$$

Additional comments

- It's useful to know that $_nC_r = {_nC_{n-r}}$. When r is a large number, the latter expression may be easier to evaluate.

- Reminder: $0! = 1$.

Detailed solutions to selected exercises

Exercise 10   Evaluate $_{25}C_{22}$.

Solution   Applying the formula $_nC_r = \dfrac{n!}{(n-r)!r!}$ with $n = 25$ and $r = 22$ yields

$$_{25}C_{22} = \frac{25!}{(25-22)!22!} = \frac{25!}{3!22!} = \frac{25 \cdot 24 \cdot 23}{3!} = 2{,}300.$$

Now do **A**.

Exercise 14   In how many ways may a student choose 1 of 4 math courses and 1 of 3 business courses? (Use the fundamental counting principle.)

Solution   There are 4 ways to choose 1 of 4 math courses and 3 ways to choose 1 of 3 business courses. Therefore, by the fundamental counting principle, there are 4×3 or 12 possibilities in all.
Now do **B**.

Exercise 16   How many distinct permutations can be made from the letters of the word NUMBER?

Solution   Since no letter is repeated in the word NUMBER, we determine the number of permutations of the 6 different letters using all of them by evaluating $n!$ with $n = 6$ to get $6! = 720$ distinct permutations.
Now do **C**.

Exercise 22   How many different 13-card bridge hands can be dealt from a deck of 52 cards? (Use the combination theorems.)

Solution   We need to find the number of combinations of 52 cards taken 13 at a time. Applying the formula $_nC_r = \dfrac{n!}{(n-r)!r!}$ with $n = 52$ and $r = 13$ gives

$$_{52}C_{13} = \frac{52!}{39!13!} = \frac{52 \cdot 51 \cdot \ldots \cdot 41 \cdot 40}{13!} = 635{,}013{,}560{,}000 \text{ different}$$

bridge hands.
Now do **D**.

Exercise 28   Find the number of different batting orders that are possible for a baseball team that starts 9 players if the pitcher must bat last. (Assume the team has only 9 players.)

Solution   In a batting order, repetition is not allowed and order is important. Thus, we use a permutations formula to fill the first 8 positions in the batting order as follows:

$$_8P_8 = 8! = 40{,}320.$$

Now do **E**.

Exercise 32   In how many ways can a 5-item multiple choice test with choices a, b, c, and d be answered?

10.6

A. Evaluate $_{15}C_3$.

B. In how many ways may a student choose 1 out of 3 available math courses and 1 out of 5 social science courses?

C. How many distinct permutations can be made from the letters of the word PRIDE?

D. In how many ways can a baseball team of 9 players be selected from among 15 candidates?

E. There is room on a shelf for displaying 6 books. How many arrangements can be made if there are 6 different books available?

F. A coin is flipped 7 times and the result of each flip is written down. How many different sequences of heads and tails are possible?

G. Solve for n: $_nC_2 = 36$.

Solution There are 4 ways to answer each of the five test items, so by the fundamental counting principle, there are

$$4 \cdot 4 \cdot 4 \cdot 4 \cdot 4 = 1{,}024 \text{ possibilities.}$$

Note that repetition is possible in the answer choices, so permutation and combination formulas are not applicable.
Now do **F**.

Exercise 42 Solve for n: $_nC_2 = 66$.

Solution $_nC_2 = \dfrac{_nP_2}{2!} = \dfrac{n(n-1)}{2} = 66$. Then $n(n-1) = 132$, so $n^2 - n - 132 = 0$, so $(n-12)(n+11) = 0$. Setting each factor equal to 0 gives $n = 12$ or $n = -11$. We reject $n = -11$, since $n \geq 2$. Thus, $n = 12$. Now do **G**.

10.7 Probability

Objectives checklist

Can you:

a. Determine the probability of a single event?

b. Determine the probability that an event will not occur given the probability that it will occur?

c. Determine the probability of event A and event B both occurring if A and B are independent?

d. Determine the probability of event A or event B occurring in the case where A and B are mutually exclusive as well as when they are not?

e. Solve applied probability problems by employing the appropriate formula(s), the definition of probability, and appropriate counting techniques as needed?

Key terms

Theoretical probability
Independent events

Mutually exclusive events

Key rules and formulas

- *Definition of probability of an event:* If an event E can occur in S ways out of a total of T equally likely outcomes, then the probability of E is

$$P(E) = \frac{S}{T} = \frac{\text{number of ways } E \text{ can occur}}{\text{total number of equally likely outcomes}}.$$

- The probability of an event is a number between 0 and 1, inclusive. If E cannot occur, $P(E) = 0$, and if E is certain, then $P(E) = 1$.

- The probability that an event E will not occur is

$$P(\text{not } E) = 1 - P(E).$$

- If A and B are independent events, then

$$P(A \text{ and } B) = P(A) \cdot P(B).$$

This rule extends to three or more independent events.

- For any events A and B,

$$P(A \text{ or } B) = P(A) + P(B) - P(A \text{ and } B).$$

If A and B are mutually exclusive, this formula reduces to

$$P(A \text{ or } B) = P(A) + P(B).$$

Additional comments

- Fractional answers are generally simplified (if possible). Probabilities may also take the form of decimals and percents.

Detailed solutions to selected exercises

Exercise 4   See text for question.

Solution

a. Since 10% of the students in the group are seniors, $P(\text{senior}) = 10\%$ or 0.1 or $\frac{1}{10}$.

b. From part **a,** $P(\text{senior}) = 0.1$. Then

$$P(\text{not senior}) = 1 - P(\text{senior}) = 0.9.$$

c. The student selected cannot be both a freshman and a sophomore simultaneously. So the events are mutually exclusive and we use the simplified "or" formula.

$$P(\text{freshman or sophomore}) = P(\text{freshman}) + P(\text{sophomore})$$

Substituting $P(\text{freshman}) = 30\%$ or 0.30 and $P(\text{sophomore}) = 25\%$ or 0.25 gives

$$P(\text{freshman or sophomore}) = 0.30 + 0.25 = 0.55.$$

d. "Neither a freshman nor a junior" is equivalent to "a sophomore or a senior," which are mutually exclusive events. So

$$P(\text{sophomore or senior}) = P(\text{sophomore}) + P(\text{senior})$$
$$= 0.25 + 0.10 = 0.35.$$

Now do **A.**

Exercise 10   What is the probability that on 4 flips of a coin the result will be 4 tails?

Solution   There are 2 equally likely outcomes on each coin flip, so $P(\text{tail}) = \frac{1}{2}$ for each flip. Furthermore, the outcomes of the flips are independent events. Thus, $P(\text{all tails}) = \frac{1}{2} \cdot \frac{1}{2} \cdot \frac{1}{2} \cdot \frac{1}{2} = \frac{1}{16}$.
Now do **B.**

Exercise 12   From the letters a, b, c, and d, three letters are selected at random without replacement. What is the probability that a person will draw the word "bad" in the correct order of spelling?

Solution   There are $_4P_3 = 24$ equally likely possible selections of 3 letters in which the order of selection is important. Of these selections, only one spells the word "bad." Thus, the probability is $\frac{1}{24}$.
Now do **C.**

10.7

A. A student is selected at random from a group in which 20 percent are freshmen, 40 percent are sophomores, 10 percent are juniors, and 30 percent are seniors. Find the probability of selecting each of the following:
   a. a sophomore
   b. not a sophomore
   c. a junior or a senior
   d. neither a junior nor a freshman

B. What is the probability that on 3 rolls of an ordinary die the result will be 3 sixes?

C. From the digits 5, 6, 7, 8, and 9 three digits are selected at random without replacement and are written down, left to right, in the order in which they are selected. What is the probability that the number 758 will be written?

D. If 10 percent of the population has type B blood, then find the following probabilities for 2 blood donors chosen at random.
  a. Both are type B.
  b. Neither is type B.
  c. 1 out of 2 is type B.

Exercise 20 If 40% of the population has type A blood, then find the following probabilities for two blood donors chosen at random.

a. Both are type A.

b. Neither is type A.

c. 1 out of 2 is type A.

Solution

a. The probability that each donor has type A blood is 0.40. Also, the blood types of the donors are independent of each other. So,

$$P(\text{both type A}) = P(\text{first type A and second type A})$$
$$= (0.40)(0.40) = 0.16.$$

b. For each donor, $P(\text{not type A}) = 1 - P(\text{type A}) = 1 - 0.40 = 0.60$. Then

$$P(\text{neither type A}) = P(\text{first not type A and second not type A})$$
$$= (0.60)(0.60) = 0.36.$$

c. Let event E be "1 out of 2 donors is type A." Then

$$P(E) = P(\text{type A, then not type A or not type A, then type A})$$
$$= (0.40)(0.60) + (0.60)(0.40)$$
$$= 0.48.$$

Now do **D.**

Sample test questions: Chapter 10

1. Complete the statement.

   a. The second term of $(x - 5y)^{15}$ is _____ .

   b. The general term a_n for the arithmetic sequence 1.1, 1.4, 1.7, . . . is _____ .

   c. The first five terms in the Fibonacci sequence are _____ .

   d. An infinite geometric series converges if the common ratio is between _____ .

   e. The expression 5! simplifies to _____ .

   f. In expanded form $\sum\limits_{i=2}^{4} 3^i$ is written as _____ .

   g. As the ratio of two integers, the repeating decimal $0.0\overline{9}$ equals _____ .

   h. A sequence of numbers in which each number after the first is found by multiplying the preceding term by a constant is called a _____ .

   i. If $a_n = (-1)^n/n^2$, then a_6 equals _____ .

   j. In sigma notation $1 + x + x^2 + x^3$ may be written as _____ .

2. Expand $(2x - 3y)^4$ by using the binomial theorem.

3. Find the sum of each of the following:

   a. First 10 terms of $1, 1.06, (1.06)^2, \ldots$

   b. First 10 terms of $1, 1.06, 1.12, \ldots$

   c. All terms of $1, 0.9, 0.81, \ldots$

   d. $\sum_{i=1}^{6} 2$

4. What is the sum of the first 100 positive integers?

5. Determine a_2 so that the following sequence is (a) an arithmetic progression and (b) a geometric progression: $1, a_2, 9, \ldots$.

6. In how many ways can 5 people line up at a supermarket checkout counter?

7. How many 3-person subcommittees can be chosen from a committee of eight people?

8. An ice cream store has 32 flavors of ice cream and 6 different toppings. How many ice cream sundaes are possible using 1 flavor of ice cream and 2 different toppings?

9. How many 5-card hands dealt from a regular deck of cards will have 3 aces and 2 kings?

10. By mathematical induction, prove that

$$\frac{1}{1 \cdot 3} + \frac{1}{3 \cdot 5} + \frac{1}{5 \cdot 7} + \cdots + \frac{1}{(2n-1)(2n+1)} = \frac{n}{2n+1}.$$

11. If a student guesses at the answers to 2 items on a multiple choice test that has 5 answer choices for each question, what is the probability that the student answers

    a. both items correctly?

    b. neither item correctly?

    c. one out of two items correctly?

12. Select the choice that completes the statement or answers the question.

    a. The sequence $1, \frac{1}{2}, \frac{1}{3}, \frac{1}{4}, \ldots$ is
       (a) an arithmetic progression
       (b) a geometric progression
       (c) neither an arithmetic nor geometric progression

    b. Does $\sum_{i=1}^{n} (-1)^{i+1}$ equal $\sum_{i=0}^{n} (-1)^i$?
       (a) Yes   (b) No

    c. If the first three terms in a geometric progression are a, a^x, a^9, then x equals
       (a) 5   (b) 3   (c) $3\sqrt{3}$ (d) $\sqrt{10}$ (e) $\frac{9}{2}$

186

d. If in an arithmetic progression, $a_3 = -20$ and $a_{35} = 28$, then the common difference is

   (a) $\frac{3}{4}$ (b) $\frac{2}{3}$ (c) $\frac{4}{3}$ (d) $\frac{1}{3}$ (e) $\frac{3}{2}$

e. The number of terms in the expansion of $(x + h)^6$ is

   (a) 2   (b) 3   (c) 5   (d) 6   (e) 7

f. Which number cannot be a probability?

   (a) 0   (b) -1 (c) 1   (d) 0.75   (e) $\frac{37}{41}$

g. The value of $_6C_4$ is

   (a) 15   (b) 30   (c) 120   (d) $\frac{3}{2}$ (e) 360

h. The number of permutations of 8 things taken 3 at a time is

   (a) 24   (b) 56   (c) 20,160   (d) 336   (e) 6,720

i. If $_nC_{10} = {}_nC_2$, then n equals

   (a) 90   (b) 45   (c) 12   (d) 8   (e) 20

Text Overview

Section	Key Concepts to Review
1.1	• Definitions of integers, rational numbers, irrational numbers, real numbers, $a < b$, $a > b$, $a \leq b$, absolute value, subtraction, factor, power, exponent, and division • Relationships among the various sets of numbers • Statements of basic properties of real numbers • Methods to add, subtract, multiply, and divide real numbers • Order of operations
1.2	• Definitions of variable, constant, algebraic expression, terms, (numerical) coefficient, and similar terms • Methods to evaluate an algebraic expression • Methods to add or subtract algebraic expressions • Guidelines for using a scientific calculator • Laws of exponents (m and n denote integers)

$$\textbf{1.}\ a^m \cdot a^n = a^{m+n} \qquad \textbf{2.}\ (a^m)^n = a^{mn} \qquad \textbf{3.}\ (ab)^n = a^n b^n \qquad \textbf{4.}\ \left(\frac{a}{b}\right)^n = \frac{a^n}{b^n} \quad (b \neq 0)$$

$$\textbf{5.}\ \frac{a^m}{a^n} = a^{m-n} = \frac{1}{a^{n-m}} \quad (a \neq 0) \qquad \textbf{6.}\ a^0 = 1 \quad (a \neq 0) \qquad \textbf{7.}\ a^{-n} = \frac{1}{a^n} \quad (a \neq 0)$$

Section	Key Concepts to Review
1.3	• Definitions of polynomial, monomial, binomial, trinomial, degree of a monomial, and degree of a polynomial • Methods to multiply various types of algebraic expressions • FOIL multiplication method • Methods to factor an expression that contains a factor common to each term, that is the difference of squares, that is a trinomial, or that is the sum or difference of cubes • Method to determine if $ax^2 + bx + c$ can be factored with integer coefficients • Factoring and product models
1.4	• Definitions of LCD and complex fractions • Methods to simplify, multiply, divide, add, and subtract fractions; to find the LCD; and to simplify a complex fraction • Equality principle: $\dfrac{a}{b} = \dfrac{c}{d}$ if and only if $ad = bc$ $(b, d \neq 0)$ • Fundamental principle: $\dfrac{ak}{bk} = \dfrac{a}{b}$ $(b, k \neq 0)$ • Operations principles $(b, d, c/d \neq 0)$

$$\frac{a}{b} \cdot \frac{c}{d} = \frac{ac}{bd} \qquad\qquad \frac{a}{b} + \frac{c}{b} = \frac{a \pm c}{b}$$

$$\frac{a}{b} \div \frac{c}{d} = \frac{a}{b} \cdot \frac{d}{c} \qquad\qquad \frac{a}{b} \pm \frac{c}{d} = \frac{ad \pm bc}{bd}$$

Section	Key Concepts to Review
1.5	• Definitions of $a^{1/n}$, index and radicand of a radical, principal square root, and nth root of a number • If m/n represents a reduced fraction such that $a^{1/n}$ represents a real number, then $\quad a^{m/n} = (a^{1/n})^m = (a^m)^{1/n}$, or equivalently, $a^{m/n} = (\sqrt[n]{a})^m = \sqrt[n]{a^m}$

Section	Key Concepts to Review
1.6	• Definitions of similar radicals and conjugates • Methods to simplify, add, subtract, multiply, and divide radicals • Methods to rationalize the denominator or the numerator • Properties of radicals (a, b, $\sqrt[n]{a}$, $\sqrt[n]{b}$ denote real numbers) **1.** $(\sqrt[n]{a})^n = a$ **2.** $\sqrt[n]{a} \cdot \sqrt[n]{b} = \sqrt[n]{ab}$ **3.** $\dfrac{\sqrt[n]{a}}{\sqrt[n]{b}} = \sqrt[n]{\dfrac{a}{b}}$ $(b \neq 0)$ **4.** $\sqrt[n]{a^n} = \begin{cases} a, \text{ if } n \text{ is odd} \\ |a|, \text{ if } n \text{ is even} \end{cases}$
2.1	• Definitions of equation, conditional equation, identity, equivalent equation, solution set, linear equation, and proportion • Methods to obtain an equivalent equation or formula • Methods to solve an equation containing fractions • Guidelines for setting up and solving word problems
2.2	• Definitions of imaginary number, complex number, equality for complex numbers, and the conjugate of a complex number • Methods to add, subtract, multiply, and divide complex numbers • Properties of conjugates • Relationships among the various sets of numbers • $i = \sqrt{-1}$ and $i^2 = -1$
2.3	• Definition of a second-degree or quadratic equation • Zero product principle: $a \cdot b = 0$ if and only if $a = 0$ or $b = 0$ • Square root property: $x^2 = n$ implies $x = \sqrt{n}$ or $x = -\sqrt{n}$ • Methods to solve a quadratic equation by the factoring method, by the square root property, by completing the square, and by using the quadratic formula • Quadratic formula: If $ax^2 + bx + c = 0$ and $a \neq 0$, then $$x = \frac{-b \pm \sqrt{b^2 - 4ac}}{2a}.$$ • Method to determine the nature of the solutions to a quadratic equation from the discriminant $(b^2 - 4ac)$
2.4	• Definitions of radical equation and equations with quadratic form • Principle of powers • Methods to solve radical equations, equations with quadratic form, and higher degree polynomial equations that factor into linear and/or quadratic factors
2.5	• Definitions of intervals and zero point • Properties of inequalities • Methods to write intervals of numbers using inequalities or interval notation • Sign rule • Method to solve certain types of inequalities by factoring and using the sign rule

Section	Key Concepts to Review
2.6	• Definition of absolute value
	• Methods to solve absolute value equations and inequalities
	• If $b \geq 0$, then $\lvert a \rvert = b$ is equivalent to $a = b$ or $a = -b$.
	• If $b > 0$, then $\lvert a \rvert < b$ is equivalent to $-b < a < b$.
	• If $b > 0$, then $\lvert a \rvert > b$ is equivalent to $a > b$ or $a < -b$.
	• Properties of absolute value
	1. $\lvert a + b \rvert \leq \lvert a \rvert + \lvert b \rvert$ **2.** $\lvert a - b \rvert \geq \lvert a \rvert - \lvert b \rvert$
	3. $\lvert ab \rvert = \lvert a \rvert \, \lvert b \rvert$ **4.** $\left\lvert \dfrac{a}{b} \right\rvert = \dfrac{\lvert a \rvert}{\lvert b \rvert}$ $(b \neq 0)$
3.1	• Rule definitions of a function and its domain and range
	• Ordered pair definitions of a function and its domain and range
	• The term $f(x)$ is read "f of x" or "f at x" and means the value of the function (the y value) corresponding to the value of x.
	• If a is in the domain of f, then ordered pairs for the function defined by $y = f(x)$ all have the form $(a, f(a))$. *Note: a is an x value and $f(a)$ is a y value.*
3.2	• Definition of the graph of a function
	• Distance formula: $d = \sqrt{(x_2 - x_1)^2 + (y_2 - y_1)^2}$ (for any two points)
	• Statement of the fundamental principle in coordinate (or analytic) geometry
	• Vertical line test
	• Basic graphs: Let a, b, c, m, and r be real number constants with $a \neq 0$.
	$y = mx + b$: straight line $y = ax^2 + bx + c$: parabola
	$y = c$: horizontal line $x^2 + y^2 = r^2$: circle (center $(0,0)$ radius r)
	$y = \lvert x \rvert$: $\vee$ shape
3.3	• Definitions of an even function and an odd function
	• The graph of an even function is symmetric about the y-axis.
	• The graph of an odd function is symmetric about the origin.
	• Methods to graph variations of a familiar function by using vertical and horizontal shifts, reflecting, stretching, and shrinking
3.4	• Definitions of $f + g$, $f - g$, $f \cdot g$, and f/g for two functions f and g
	• The symbol "$\circ$" denotes the operation of composition. The composite functions are $(f \circ g)(x) = f[g(x)]$ and $(g \circ f)(x) = g[f(x)]$.
	• Methods to determine the domain of $f + g$, $f - g$, $f \cdot g$, f/g, $f \circ g$, $g \circ f$
3.5	• Definitions of a one-to-one function and inverse functions
	• The special symbol f^{-1} is used to denote the inverse function of f.
	• Methods to determine if the inverse of a function is a function and how to find f^{-1}, if it exists.
	• Horizontal line test
	• f and f^{-1} interchange their domain and range.
	• The graphs of f and f^{-1} are reflections of each other across the line $y = x$.

Section	Key Concepts to Review
3.6	• Slope formula: $m = \dfrac{\Delta y}{\Delta x} = \dfrac{y_2 - y_1}{x_2 - x_1} \qquad (x_2 \neq x_1)$
	• Average rate of change from x to $x + h$: $\dfrac{\Delta y}{\Delta x} = \dfrac{f(x + h) - f(x)}{h}, h \neq 0$
3.7	• The statement "y varies directly as x" means there is some positive number k (variation constant) such that $y = kx$.
	• The statement "y varies inversely as x" means there is some positive number k such that $xy = k$ or $y = k/x$.
4.1	• Definitions of polynomial function and linear function
	• Point-slope equation: $y - y_1 = m(x - x_1)$
	• Slope-intercept equation: $y = mx + b$
	• For two lines with slopes m_1 and m_2: parallel lines: $m_1 = m_2$ perpendicular lines: $m_1 \cdot m_2 = -1$
4.2	• Definitions of quadratic function, quadratic inequality, and axis of symmetry
	• Methods to graph a quadratic function
	• Axis of symmetry formula: $x = -b/2a$
	• Vertex formula: $(-b/2a, f(-b/2a))$
	• The graph of $f(x) = a(x - h)^2 + k$ (with $a \neq 0$) is the graph of $y = ax^2$ (a parabola) shifted so the vertex is (h,k) and the axis of symmetry is $x = h$.
	• Methods to solve a quadratic inequality
4.3	• Long division procedure for the division of polynomials
	• Synthetic division procedure for the division of a polynomial by $x - b$
	• Remainder theorem
4.4	• Definitions of a zero of a function and the multiplicity of a zero
	• Factor theorem
	• Fundamental theorem of algebra
	• Number of zeros theorem
	• Theorem about when complex zeros come in conjugate pairs
	• Theorem about when irrational zeros of the form $a \pm b\sqrt{c}$ come in conjugate pairs
	• Theorem about graphing polynomial functions by knowing the multiplicity of real number zeros
4.5	• Rational zero theorem
	• Descartes' rule of signs
	• Upper and lower bounds theorem
	• Location theorem
4.6	• Definition of a rational function
	• Methods to determine vertical, horizontal, or oblique asymptotes
	• Horizontal asymptote theorem
	• Theorem about the behavior of a graph near any vertical asymptote(s)
	• Outline of the procedure for graphing rational functions

Section	Key Concepts to Review

5.1	• Definition of the exponential function with base b
	
	• For $b > 0$, $b \neq 1$, and any real number x, if x lies between the rational numbers r and s, then b^x lies between b^r and b^s.
	
	• All previous laws of exponents hold for real number exponents.
	
	• For $f(x) = b^x$ with $b > 0$, $b \neq 1$:
	  Domain: $(-\infty, \infty)$ Horizontal asymptote: x-axis
	  Range: $(0, \infty)$ y-intercept: $(0,1)$
	
	• If $b > 0$, $b \neq 1$, then $b^x = b^y$ implies $x = y$.
	
	• Methods from Section 3.3 to graph variations of $f(x) = b^x$

5.2	• Definitions of a logarithm, a common logarithm, and the logarithmic function with base b
	
	• $\log_b N = L$ is equivalent to $b^L = N$.
	
	• For $b > 0$, $b \neq 1$, and $x > 0$,
	    $y = \log_b x$ if and only if $x = b^y$
	and for $f(x) = \log_b x$:
	  Domain: $(0, \infty)$ Vertical asymptote: y-axis
	  Range: $(-\infty, \infty)$ x-intercept: $(1,0)$
	
	• The logarithmic function $y = \log_b x$ and the exponential function $y = b^x$ are inverse functions. Consequently,
	  a. The functions interchange their domain and range.
	  b. The graphs of the functions are symmetric about the line $y = x$.
	  c. Calculator evaluations of exponential and log expressions often involve the $\boxed{\text{INV}}$ key.
	
	• Methods from Section 3.3 to graph variations of $f(x) = \log_b x$

5.3	• Properties of logarithms (for b, x, $y > 0$, $b \neq 1$, and k any real number):
	
	  1. $\log_b xy = \log_b x + \log_b y$ **2.** $\log_b(x/y) = \log_b x - \log_b y$
	  3. $\log_b x^k = k \log_b x$ **4.** $\log_b b = 1$
	  5. $\log_b 1 = 0$ **6.** $\log_b b^x = x$
	  7. $b^{\log_b x} = x$

5.4	• If x, y, $b > 0$, with $b \neq 1$, then $x = y$ implies $\log_b x = \log_b y$; and $\log_b x = \log_b y$ implies $x = y$. • Change of base formula: $\log_b x = \dfrac{\log_a x}{\log_a b}$ • Methods to solve exponential and logarithmic equations
5.5	• Definition of a natural logarithm • As n gets larger and larger, $(1 + 1/n)^n$ gets closer and closer to an irrational number that is denoted by the letter e. To six significant digits, $e \approx 2.71828$. • Compound interest formula: $A = P\left(1 + \dfrac{r}{n}\right)^{nt}$ • Continuous growth or decay formula: $A = A_0 e^{kt}$ • $y = \ln x$ and $y = e^x$ are inverse functions. • Graphs of $y = \ln x$ and $y = e^x$ • $\ln e^x = x$ and $e^{\ln x} = x$ (if $x > 0$)

Section	Key Concepts to Review
6.1	• Definitions of right triangle, acute angle, hypotenuse, similar triangles, cofunctions, complementary angles, one minute (written 1′), reciprocal functions, angle of elevation, and angle of depression
	• Definition of the trigonometric functions of an acute angle θ of a right triangle
	• The side lengths in a 30–60–90 triangle are in the ratio of $1:\sqrt{3}:2$.
	• The side lengths in a 45–45–90 triangle are in the ratio of $1:1:\sqrt{2}$.
	• A trigonometric function of any acute angle is equal to the corresponding cofunction of the complementary angle.
	• Guidelines for using a scientific calculator or Table 5 to evaluate a trigonometric function of an acute angle
	• Methods to solve a right triangle
	• Guidelines for accuracy in computed results
6.2	• Definitions of ray, measure of an angle, standard position of angle θ, quadrantal angle, coterminal angle, and reference angle
	• Definition of the trigonometric functions of an angle in standard position
	• Chart summarizing the signs of the trigonometric ratios
	• Methods to determine (if defined) approximate trigonometric values for any angle and exact values in special cases
6.3	• Definition of 1 radian
	• Radian measure formula: $\theta = s/r$
	• Degrees to radians formula: $1° = \pi/180$ radians
	• Radians to degrees formula: 1 radian $= 180°/\pi$
	• Area formula: $A = \frac{1}{2}r^2\theta$ (θ in radians)
	• Formula relating linear velocity (v) and angular velocity (ω): $v = \omega r$
6.4	• Definition of the trigonometric functions for a unit circle
	• Definitions of trigonometric identity and reference number
	• In a unit circle the same real number measures both the central angle θ and the intercepted arc s. (That is, $\theta = s$.)
	• For the sine and cosine functions, the domain is the set of all real numbers, and the range is the set of real numbers between -1 and 1, inclusive.
	• Methods to determine (if defined) approximate trigonometric values for any number and exact values in special cases
	• For any trigonometric function f, we have $f(s + 2\pi k) = f(s)$, where k is an integer.
	• Negative angle identities
	• When using a scientific calculator, be sure to set the calcuator for radian mode.
6.5	• Definitions of periodic function, period, amplitude, and phase shift
	• For $y = a\sin(bx + c)$ and $y = a\cos(bx + c)$, with $b > 0$, Amplitude $= |a|$, Period $= 2\pi/b$, Phase shift $= -c/b$
6.6	• Graphs of $y = \tan x$, $y = \cot x$, $y = \sec x$, and $y = \csc x$
	• Domain, range, and period for $y = \tan x$, $y = \cot x$, $y = \sec x$, and $y = \csc x$
	• Relations between a function and its reciprocal function

Section	Key Concepts to Review
6.7	• Statements of the eight *fundamental identities* (See back endpaper 1–8.) • Guidelines for proving identities
6.8	• Statements of the sum, difference, double-angle, and half-angle formulas for sine and cosine • Verification of the tangent formulas and the reduction formulas by using the above identities • Summary of the major trigonometric identities
6.9	• Methods to solve certain trigonometric equations (Reference numbers, identities, and a scientific calculator may be involved.) • Determine solutions between 0 and 2π as follows:

Quadrant	Solution
1	Reference number
2	π − reference number
3	π + reference number
4	2π − reference number

	• We generate all the solutions to a trigonometric equation by adding multiples of 2π to the solutions that are in the interval $[0,2\pi)$.
6.10	• The inverse sine function is denoted by arcsin or $\sin^{-1}$. By definition, $y = \arcsin x$ if and only if $x = \sin y$, where $-1 \leq x \leq 1$ and $-\pi/2 \leq y \leq \pi/2$. Similar remarks hold for the other inverse trigonometric functions. • Domain and range of the six inverse trigonometric functions • Graphs of $y = \arcsin x$, $y = \arccos x$, and $y = \arctan x$
7.1	• Law of sines: $\dfrac{\sin A}{a} = \dfrac{\sin B}{b} = \dfrac{\sin C}{c}$ • Guidelines on when to use the law of sines • When given the measures for two sides of a triangle and the angle opposite one of them, there may be one triangle that fits the data (see Example 3) or there may be two triangles that fit the data (see Example 4). Sometimes no triangle can be constructed from the data; then we say the data is inconsistent.
7.2	• Law of cosines: $a^2 = b^2 + c^2 - 2bc \cos A$ $b^2 = a^2 + c^2 - 2ac \cos B$ $c^2 = a^2 + b^2 - 2ab \cos C$ • Guidelines on when to use the law of cosines
7.3	• Definitions of vector, resultant of two (or more) vectors, and components of a vector • Methods to determine the resultant of two (or more) vectors • Methods to determine the horizontal and vertical components of a vector • Resultant formulas: $R = \sqrt{(R_x)^2 + (R_y)^2}$ and $\tan \theta = R_y/R_x$

Section	Key Concepts to Review
7.4	• Definitions of complex plane, real axis, imaginary axis, and the argument and absolute value of a complex number • Graph of a complex number • In trigonometric form $a + bi$ is written as $r(\cos \theta + i \sin \theta)$. • The absolute value, r, is given by $r = \sqrt{a^2 + b^2}$. • The argument, θ, is given by $\tan \theta = b/a$. • Methods to multiply and divide two complex numbers in trigonometric form • De Moivre's theorem
7.5	• Definitions of pole, polar axis, and polar coordinates • If $r < 0$, we plot (r,θ) by measuring $|r|$ units in a direction opposite to the terminal ray of θ. • Formulas for different representations of a given point: $\quad (r,\theta) = (r,\theta + n \cdot 360°)$, where n is an integer $\quad (-r,\theta) = (r,\theta + n \cdot 180°)$, where n is an odd integer • Two tests for x- and y-axis symmetry • Basic graphs: $\quad r = a, r = \pm 2a \sin \theta, r = \pm 2a \cos \theta$: circle $\quad r = a \pm a \sin \theta, r = a \pm a \cos \theta$: cardioid $\quad r = a \sin n\theta, r = a \cos n\theta$: rose $\begin{cases} n \text{ loops, if } n \text{ is odd} \\ 2n \text{ loops, if } n \text{ is even} \end{cases}$ • Formulas relating rectangular coordinates and polar coordinates: $\quad x = r \cos \theta, y = r \sin \theta, x^2 + y^2 = r^2, \tan \theta = y/x$
8.1	• Fundamental relationship between an equation and its graph: Every ordered pair that satisfies the equation corresponds to a point in its graph, and every point in the graph corresponds to an ordered pair that satisfies the equation. • Procedures for finding an equation that corresponds to a given geometric condition • Procedures for proving certain theorems in plane geometry by using a coordinate system and algebraic methods • Midpoint formula: $\left(\dfrac{x_1 + x_2}{2}, \dfrac{y_1 + y_2}{2}\right)$
8.2	• Definition of circle • Standard form of a circle of radius r with center (h,k): $\quad (x - h)^2 + (y - k)^2 = r^2$

Section	Key Concepts to Review
8.3	• Definitions of ellipse, foci, major axis, minor axis, and vertices
	• Summary for an ellipse:

Foci	Center	Standard Form ($a > b$)
On x-axis at $(\pm c, 0)$	Origin	$\dfrac{x^2}{a^2} + \dfrac{y^2}{b^2} = 1$
On y-axis at $(0, \pm c)$	Origin	$\dfrac{x^2}{b^2} + \dfrac{y^2}{a^2} = 1$
On major axis parallel to x-axis	(h,k)	$\dfrac{(x-h)^2}{a^2} + \dfrac{(y-k)^2}{b^2} = 1$
On major axis parallel to y-axis	(h,k)	$\dfrac{(x-h)^2}{b^2} + \dfrac{(y-k)^2}{a^2} = 1$

• In general, $a^2 = b^2 + c^2$, where a, b, and c represent the following distances:

a is the distance from the center to the endpoints on the major axis
b is the distance from the center to the endpoints on the minor axis
c is the distance from the center to the foci

8.4
• Definitions of hyperbola, foci, transverse axis, vertices, conjugate axis, and asymptotes of the hyperbola
• Summary for a hyperbola:

Foci	Center	Standard Form	Asymptotes
On x-axis at $(\pm c, 0)$	Origin	$\dfrac{x^2}{a^2} - \dfrac{y^2}{b^2} = 1$	$y = \pm\dfrac{b}{a}x$
On y-axis at $(0, \pm c)$	Origin	$\dfrac{y^2}{a^2} - \dfrac{x^2}{b^2} = 1$	$y = \pm\dfrac{a}{b}x$
On transverse axis parallel to x-axis	(h,k)	$\dfrac{(x-h)^2}{a^2} - \dfrac{(y-k)^2}{b^2} = 1$	$y - k = \pm\dfrac{b}{a}(x-h)$
On transverse axis parallel to y-axis	(h,k)	$\dfrac{(y-k)^2}{a^2} - \dfrac{(x-h)^2}{b^2} = 1$	$y - k = \pm\dfrac{a}{b}(x-h)$

• In general, $c^2 = a^2 + b^2$, where a, b, and c represent the following distances:

a is the distance from the center to the endpoints on the transverse axis
b is the distance from the center to the endpoints on the conjugate axis
c is the distance from the center to the foci

Section	Key Concepts to Review

8.5
- Definitions of parabola, directrix, focus, axis of symmetry, and vertex
- Summary for a parabola with vertex at the origin: (In all cases p gives the distance from the vertex to the focus and from the vertex to the directrix. See Figures 8.24–8.27.)

Standard Form	Opens	Axis of Symmetry	Focus	Directrix
$y^2 = 4px$	Right	x-axis	$(p,0)$	$x = -p$
$y^2 = -4px$	Left	x-axis	$(-p,0)$	$x = p$
$x^2 = 4py$	Upward	y-axis	$(0,p)$	$y = -p$
$x^2 = -4py$	Downward	y-axis	$(0,-p)$	$y = p$

- Figure 8.29 summarizes the cases when the vertex of the parabola is at the point (h,k) and the directrix is parallel to the x- or y-axis.

8.6
- The general form of an equation of a conic section with axis or axes parallel to the coordinate axes is
$$Ax^2 + Cy^2 + Dx + Ey + F = 0,$$
where A and C are not both zero.
- Chart summarizing the graphing possibilities for the above equation

9.1
- Definitions of linear equations with n variables, linear system, inconsistent system, and dependent system
- The solution set of a system of linear equations in two variables is the set of all the ordered pairs that satisfy both equations. Graphically this corresponds to the collection of points where the lines intersect.
- Methods to solve a system of equations by the substitution method and by the addition-elimination method

9.2
- Definitions of determinant, minor of an element, and cofactor of an element
- To evaluate a 2 by 2 determinant, use $\begin{vmatrix} a_1 & b_1 \\ a_2 & b_2 \end{vmatrix} = a_1b_2 - b_1a_2$.
- Cramer's rule
- Method to evaluate 3 by 3 (and more complicated) determinants

9.3
- Definitions of triangular form, equivalent systems, matrix, entry or element of a matrix, coefficient matrix, and augmented matrix
- The following operations are used in Gaussian elimination to solve a linear system.

Elementary Operations on Equations	Elementary Row Operations on Matrices
1. Multiply both sides of an equation by a nonzero number.	1. Multiply each entry in a row by a nonzero number.
2. Add a multiple of one equation to another.	2. Add a multiple of the entries in one row to another row.
3. Interchange two equations.	3. Interchange two rows.

Section	Key Concepts to Review
9.4	• Definitions of dimension, square matrix, zero matrix, equal matrices, scalar, $n \times n$ identity matrix, and the multiplicative inverse of A (symbolized A^{-1})
	• Definitions for scalar multiplication and addition, subtraction, and multiplication of matrices
	• Six basic properties of matrices
	• Only matrices of the same dimension can be added or subtracted.
	• The product AB of two matrices is defined only when the number of columns in A matches the number of rows in B.
	• Methods to determine if A^{-1} exists, and if it does, how to find it
	• If A^{-1} exists, the solution to the linear system $AX = B$ is given by $X = A^{-1}B$.
9.5	• Definitions of partial fraction decomposition, partial fraction, and irreducible quadratic factor
	• Methods to determine the partial fraction decomposition of $P(x)/Q(x)$, where $P(x)$ and $Q(x)$ are (nonzero) polynomials (Note that the denominator in each partial fraction in the decomposition is either a linear factor, a repeated linear factor, an irreducible quadratic factor, or a repeated irreducible quadratic factor.)
9.6	• Methods to solve nonlinear systems of equations by the substitution method and by the addition-elimination method
9.7	• Definitions of half-plane, vertices or corners of a region, objective function, constraints, feasible solution, and convex region
	• Methods to graph inequalities in two variables
	• Methods to solve a system of inequalities
	• The solution set of a system of inequalities is the intersection of the solution sets of all the individual inequalities in the system.
	• Methods to find the maximum and/or minimum value of an objective function subject to given constraints
	• Corner point theorem
10.1	• Definitions of sequence, terms (of the sequence), arithmetic progression, common difference, geometric progression, common ratio and Fibonacci sequence
	• Formulas for the nth term:
	$\quad$ arithmetic progression: $a_n = a_1 + (n-1)d$ $\quad$ geometric progression: $a_n = a_1 r^{n-1}$ $\quad$ Fibonacci sequence: $a_1 = 1$, $a_2 = 1$, $a_n = a_{n-1} + a_{n-2}$ for $n \geq 3$
10.2	• Definitions of series, arithmetic series, and geometric series
	• Formulas for the sum of the first n terms:
	$\quad$ arithmetic series: $S_n = \dfrac{n}{2}(a_1 + a_n)$ or $S_n = \dfrac{n}{2}[2a_1 + (n-1)d]$
	$\quad$ geometric series: $S_n = \dfrac{a_1 - a_1 r^n}{1 - r}$ or $S_n = \dfrac{a_1 - a_n r}{1 - r}$
	• The Greek letter Σ (read "sigma") is used to simplify the notation involved with series. We define sigma notation as follows:
	$$\sum_{i=1}^{n} a_i = a_1 + a_2 + \cdots + a_n.$$
10.3	• An infinite geometric series with $\lvert r \rvert < 1$ converges to the value or sum S given by $S = \dfrac{a_1}{1 - r}$.

Section	Key Concepts to Review
10.4	• Principle of mathematical induction
	• We prove by mathematical induction that a statement is true for all positive integers by doing the following.
	1. Show by direct substitution that the statement (or formula) is true for $n = 1$.
	2. Show that if the statement is true for any positive integer k, then it is also true for the next highest integer $k + 1$.
	• Two methods for establishing the "$k + 1$" statement from the assumption of the "k" statement
10.5	• Binomial theorem: For any positive integer n, $$(a + b)^n = a^n + \frac{n}{1!}a^{n-1}b + \frac{n(n-1)}{2!}a^{n-2}b^2 + \cdots$$ $$+ \frac{n(n-1)(n-2)\cdots(n-r+1)}{r!}a^{n-r}b^r + \cdots + b^n,$$ or an alternate form is $$(a + b)^n = \sum_{r=0}^{n}\binom{n}{r}a^{n-r}b^r.$$
	• The binomial coefficient $\binom{n}{r}$ is defined by $$\binom{n}{r} = \frac{n!}{r!(n-r)!}.$$
	• The rth term of the binomial expansion of $(a + b)^n$ is $$\binom{n}{r-1}a^{n-(r-1)}b^{r-1}.$$
	• The symbol $n!$ (read "n factorial") means the product $n \cdot (n-1) \cdot \cdots \cdot 3 \cdot 2 \cdot 1$.
	• Pascal's triangle
10.6	• Definitions of permutation and combination
	• Method to construct a tree diagram
	• Fundamental counting principle

	• Permutation formulas:
	   n objects, n at a time: $_nP_n = n!$
	
	   n objects, r at a time: $_nP_r = \underbrace{n(n-1)\cdots(n-r+1)}_{r \text{ factors}} = \frac{n!}{(n-r)!}$

	• Formula for distinguishable permutations (if certain objects are all alike)
	• Combination formulas: n objects, r at a time: $_nC_r = \frac{_nP_r}{r!} = \frac{n!}{(n-r)!r!}$
	• The order of the objects is important in a permutation but does not matter in a combination.
10.7	• Definitions of probability of an event, independent events, and mutually exclusive events
	• The probability of an event is a number between 0 and 1, inclusive, with impossible events assigned a probability of 0 and events that must take place assigned a probability of 1.

	• Probability formulas
	   1. $P(\text{not } E) = 1 - P(E)$
	   2. $P(A \text{ and } B) = P(A) \cdot P(B)$ (if A and B are independent events)
	   3. a. $P(A \text{ or } B) = P(A) + P(B) - P(A \text{ and } B)$ (for any events A and B)
	      b. $P(A \text{ or } B) = P(A) + P(B)$ (if A and B are mutually exclusive)

Answers to Margin Exercises

Chapter 1

Section 1.1

A. $1.1\overline{6}$ **B.** Real, irrational   **C.** Real, rational   **D.** Commutative property of addition   **E.** Associative property of multiplication
F. $<$ **G.** -4.83 **H.** 4   **I.** 51   **J.** True

Section 1.2

A. -100 **B.** $-\frac{8}{3}$ **C.** $7x^2y - 4xy + 2y^3$ **D.** $-10x + 9$ **E.** $1{,}296$ **F.** $12/x^5$ **G.** 2^{5x-2} **H.** $8y^{-2mn}$ **I.** Yes   **J.** 13 ft

Section 1.3

A. $-2y^2 + 15xy - 18x^2$ **B.** $4ab^3(a^6 - 4a^4b^6 + 6b^2)$ **C.** $(12x^3 - 11a^2)(12x^3 + 11a^2)$ **D.** $(x + 5)(x - 3)$ **E.** $(2a - d)(a + 5d)$
F. $(x + 1)(x^2 + 5x + 7)$ **G.** $(3x^2 + 2)(x - 4)$ **H.** No   **I.** $4x^3 + 6x^2h + 4xh^2 + h^3$

Section 1.4

A. $\dfrac{x + 3}{x - 4}$ **B.** $\dfrac{x(x + 1)}{x^2 + 2x + 4}$ **C.** $\dfrac{4xy + 2x - 5y}{x^2y^2}$ **D.** $\dfrac{x^2 - 5x + 15}{x^2 - 9}$ **E.** $\dfrac{x^2 - 4x + 1}{(x - 2)^3}$ **F.** $\dfrac{ax - 1}{3 + ax}$ **G.** $\dfrac{3}{(x + h + 2)(x + 2)}$
H. $\dfrac{x^2 - x - 1}{x^2 - 1}$

Section 1.5

A. -93 **B.** 9   **C.** $y^{2/3}$ **D.** $x^{11/5}$ **E.** $\sqrt[2n]{x^n + 2}$ **F.** $1/(x - 1)^{3/2}$ **G.** $(-2x^{1/5} + 7)/x^{3/5}$

Section 1.6

A. $3x^2y^9\sqrt{2xy}$ **B.** $\sqrt[3]{3xy^2}/y^3$ **C.** $12\sqrt{5} - 21\sqrt{3}$ **D.** $2x^2\sqrt[3]{x}$ **E.** $7\sqrt[3]{1 - x}$ **F.** $7 - 4\sqrt{3}$ **G.** $6x^3y^2\sqrt{x}$ **H.** $-9y$
I. $\dfrac{-(3 + 3\sqrt{5})}{4}$ **J.** $\dfrac{1 - x}{(1 + \sqrt{x})^2}$ **K.** $\dfrac{x^2 - \sqrt{xy}}{y^2}$

Chapter 2

Section 2.1

A. $\left\{\frac{11}{3}\right\}$ **B.** Set of all real numbers   **C.** $\{-14\}$ **D.** $\left\{\frac{1}{5}\right\}$ **E.** 1,927 in.²   **F.** 1 hr 12 min.   **G.** $r = (a - p)/(tp)$
H. $y = z/(xz - 2)$

Section 2.2

A. $1 + (\sqrt{2}/2)i$ **B.** $5 - 5i$ **C.** $(3i)^4 + 2(3i)^2 - 63 = 81 - 18 - 63 = 0$ **D.** Yes   **E.** $-\frac{9}{10} + \frac{3}{10}i$ **F.** $\frac{3}{25} - \frac{4}{25}i$
G. $\overline{z \cdot w} = \overline{z} \cdot \overline{w} = 6 - 2i$

Section 2.3

A. $\{\frac{1}{3}, -5\}$ **B.** $\{-13, 17\}$ **C.** $\frac{25}{16}$ **D.** $\left\{\dfrac{1 \pm \sqrt{13}}{3}\right\}$ **E.** $\left\{\dfrac{5 \pm \sqrt{65}}{4}\right\}$ **F.** Unequal, rational   **G.** 4 ft by 3 ft

Section 2.4

A. $\{6\}$ **B.** $\{-1, -2\}$ **C.** $\{4\}$ **D.** $\{\pm 0, \pm 2, \pm 4\}$

Section 2.5

A. $(-\infty, -1]$

B. $(-\infty, \infty)$

C. $[1, 4)$

D. $[-2, 3]$ **E.** $(-\infty, -6) \cup (2, \infty)$ **F.** $(-1, -\frac{2}{3})$ **G.** [72.5 in., 73.5 in.)   **H.** (0 ft, 5 ft)

Section 2.6

A. $=$ **B.** $\{0, 2\}$ **C.** $(-c - b, c - b)$ **D.** $(-\infty, -5] \cup [11, \infty)$ **E.** $(-\infty, 1) \cup (3, \infty)$ **F.** $\{x: |x| > 5\}$

Chapter 3

Section 3.1

A. No **B.** *D:* Set of all real numbers except 4; *R:* Set of all real numbers except 0 **C.** $(-1,1)$ No; $(1,1)$ No; $(-1,-1)$ Yes; $(1,-1)$ Yes
D. -3 **E.** Yes **F.** *D:* $\{5,4,3,2\}$; *R:* $\{3,2,1,0\}$ **G.** $f(-1) = -3, f(0) = -1, f(1) = 1$ **H.** $x = -3$ **I.** **(a)** 8 **(b)** 8 **(c)** Yes
J. **(a)** 5 **(b)** 5 **(c)** 10 **K.** 1 **L.** $P = 4\sqrt{A}$; *D:* $(0,\infty)$ **M.**
$$c = \begin{cases} 150, & \text{if } 0 \leq n \leq 20 \\ 150 + 5(n - 20), & \text{if } 20 < n \leq 50 \\ 300 + 4(n - 50), & \text{if } 50 < n \leq 100 \end{cases} \quad D: [0,100]$$

Section 3.2

A. $25\pi/2$ **B.** $[0,\infty)$

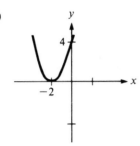

C. $(-\infty,0]$

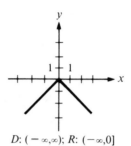

$D: (-\infty,\infty); R: (-\infty,0]$

D. **(a)** $(-\infty,\infty)$ **(b)** $[-1,1]$ **(c)** $f(0) = 0; f(3) = -1$ **(d)** $\{3\}$ **(e)** $\{-4, -2,0,2,4\}$
(f) $\{x: 0 < x < 2\}$ **(g)** $\{x: -2 < x < 0 \text{ or } 2 < x < 4\}$ **(h)** $\{x: -1 < x < 1\}$ **E.** Yes

Section 3.3

A. Odd **B.** Neither **C.**

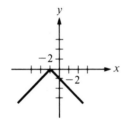

D.

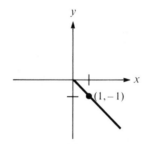

Section 3.4

A. $(f + g)(x) = x^4 + \sqrt[4]{x}$, $(f - g)(x) = x^4 - \sqrt[4]{x}$, $(f \cdot g)(x) = x^{17/4}$, $(f/g)(x) = x^{15/4}$, $(f \circ g)(x) = x$, $(g \circ f)(x) = |x|$;
$D_{f+g} = D_{f-g} = D_{f \cdot g} = D_{f \circ g} = [0,\infty)$, $D_{f/g} = (0,\infty)$, $D_{g \circ f} = (-\infty,\infty)$ **B.** $g(x) = x^3 + 9, f(x) = \sqrt[4]{x}$ **C.** $(3x + 3)^3$

Section 3.5

A. *D:* $\{-1,2,-5\}$, *R:* $\{3,4,5\}$ **B.** Yes **C.** Yes **D.**

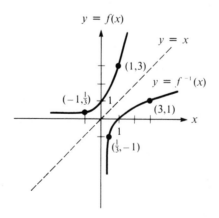

E.

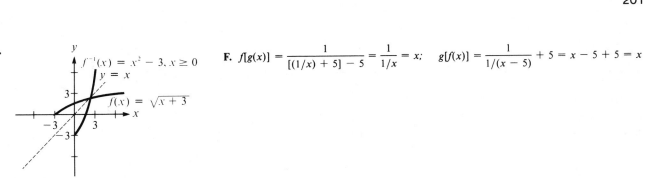

F. $f[g(x)] = \dfrac{1}{[(1/x) + 5] - 5} = \dfrac{1}{1/x} = x;$ $g[f(x)] = \dfrac{1}{1/(x - 5)} + 5 = x - 5 + 5 = x$

G. $y = \frac{5}{9}(x - 32) + 273.16$; converts degrees Fahrenheit to degrees Kelvin

Section 3.6

A. 2   **B.** $m = 1.20$; For each \$1 increase in the price a textbook costs the bookstore, there is a \$1.20 increase in the price at which a textbook is sold.   **C.** $\frac{1}{19}$ **D.** $\dfrac{1}{\sqrt{x + h + 2} + \sqrt{x + 2}}$ **E.** $19.6 + 4.9h$ m/sec; 20.09 m/sec; 19.649 m/sec; 19.6049 m/sec; 19.6 m/sec

Section 3.7

A. $\frac{36}{7}$ **B.** Multiplied by 4   **C.** Multiplied by $\frac{3}{4}$

Chapter 4

Section 4.1

A. No   **B.** $y = -\frac{2}{5}x - \frac{19}{5}$ **C.** $y = -6$ **D.** $\frac{3}{4}$; (0,2)   **E.** $y = 2x - 4$

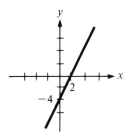

F. $f(x) = -3x - 3$ **G.** (a) $y = 3x + 1$ (b) $y = -\frac{1}{3}x - 9$ **H.** (a) $y = 9x + 50$ (b) 185 lb   (c) 50 lb   (d) 9 lb
I. $m_{AB} = m_{BC} = \frac{1}{2}$

Section 4.2

A.

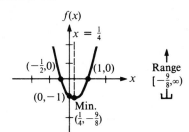

(a) x-intercepts: (1,0), $(-\frac{1}{2},0)$; y-intercept: (0,−1)   (b) $x = \frac{1}{4}$ (c) $(\frac{1}{4}, -\frac{9}{8})$, minimum
(d) $[-\frac{9}{8}, \infty)$

B. (1,5); $x = 1$ **C.** 2,500 ft²   **D.** $[-4,2]$ **E.** $(-\infty, \infty)$

Section 4.3

A. $3x^3 - x^2 + 6x - 8 = (x^2 + 2)(3x - 1) - 6$ **B.** $3x^3 + 2x^2 + 1 = (x + 2)(3x^2 - 4x + 8) - 15$ **C.** $-3, -\frac{73}{27}$

Section 4.4

A. 0, multiplicity 3; -4, multiplicity 2; 3, multiplicity 2; degree 7
B. $P(x) = x[x - (-3 + i)][x - (-3 - i)][x - (2 + \sqrt{5})][x - (2 - \sqrt{5})]$ **C.**

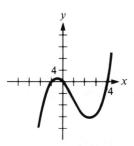

D. $P(x) = [x - (1 + 2i)][x - (1 - 2i)](x - \sqrt{3})(x + \sqrt{3})$ **E.** $2 + i, -\sqrt{2}$ **F.** $(1 + \sqrt{7}i)/4, (1 - \sqrt{7}i)/4$

Section 4.5

A. $\pm 5, \pm 1, \pm\frac{1}{2}, \pm\frac{1}{3}, \pm\frac{1}{6}, \pm\frac{5}{2}, \pm\frac{5}{3}, \pm\frac{5}{6}$ **B.** 0 positive, 1 negative   **C.** $-\frac{1}{2}, \sqrt{5}, -\sqrt{5}$ **D.** 1.6

Section 4.6

A. $x = -3, x = 8$ **B.** $VA: x = 0, HA: y = 1$

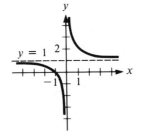

C. $VA: x = 0, OA: y = x$

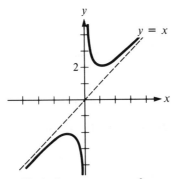

Vertical asymptote: $x = 0$
Horizontal asymptote: none

D.

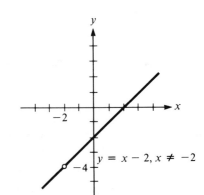

Chapter 5

Section 5.1

A. $(2,\frac{1}{9})$, $(1,\frac{1}{3})$, $(-\frac{1}{3},\sqrt[3]{3})$, $(-1,3)$ **B.**

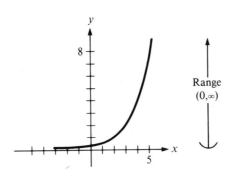

C. -2 **D.** **(a)** 13,000   **(b)** 416,000   **(c)** 6

E. $577,284.38

Section 5.2

A. $\log_{1/3} 81 = -4$ **B.** $(\frac{1}{5})^{-2} = 25$ **C.** $\frac{1}{3}$ **D.** 1,000   **E.** $2, -1, \frac{5}{2}$ **F.**

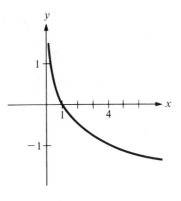

G.

y, x graph

H. $(2,\infty)$ **I.** 4.8   **J.** 3.9×10^{-8}

Section 5.3

A. $\log_b x - \frac{1}{2}\log_b y$ **B.** $\log_b(x^2/\sqrt[3]{y})$ **C.** $\frac{4}{3}$ **D.** x/y **E.** $4x + y$ **F.** Let $b = 10$, $x = 10$, $y = 10$. Then $\log_{10}(20) \neq (1)(1)$.

Section 5.4

A. $\{0.6045\}$ **B.** $\{0.1838\}$ **C.** $\{\frac{1}{2}\sqrt[3]{10}\}$ **D.** $\{250\}$ **E.** $\{1\}$ **F.** 0.8192   **G.** 1   **H.** 10.2 years

Section 5.5

A. $\{0.2369\}$ **B.** 125   **C.** $\{-0.5040\}$ **D.** $k \cdot \ln(y/c)$ **E.** **(a)** $1,816.70   **F.** 62.1 days   **G.** 2.46 hours   **H.** $t = \left(\frac{1}{a}\right)\ln\left(\frac{ry-y}{c}\right)$

Section 6.1

A. $\sin X = x/y$, $\csc X = y/x$, $\cos X = z/y$, $\sec X = y/z$, $\tan X = x/z$, $\cot X = z/x$, $\sin Z = z/y$, $\csc Z = y/z$, $\cos Z = x/y$, $\sec Z = y/x$, $\tan Z = z/x$, $\cot Z = x/z$ **B.** $\cos \alpha$ or $\sin \theta$ **C.** 4/3   **D.** $\sin \theta = 1/\sqrt{2}$, $\cos \theta = 1/\sqrt{2}$, $\sec \theta = \sqrt{2}$, $\tan \theta = 1$, $\cot \theta = 1$ **E.** $\sec 59.5°$
F. No   **G.** 1.236   **H.** 43.0° or 43°00′   **I.** $b = 6.71$ ft, $A = 41.8°$ or 41°50′, $B = 48.2°$ or 48°10′   **J.** 13 feet   **K.** 60°

Section 6.2

A. $\sin \theta = -4/\sqrt{17}$, $\cos \theta = -1/\sqrt{17}$, $\tan \theta = 4$, $\csc \theta = -\sqrt{17}/4$, $\sec \theta = -\sqrt{17}$, $\cot \theta = 1/4$ **B.** $\sin \theta = \sqrt{3}/2$, $\cos \theta = 1/2$,
$\tan \theta = \sqrt{3}$, $\csc \theta = 2/\sqrt{3}$, $\cot \theta = 1/\sqrt{3}$ **C.** 50°   **D.** 1   **E.** $\sqrt{2}$ **F.** −2.145

Section 6.3

A. 36.9 ft   **B.** 20 ft²   **C.** $23\pi/12$ **D.** 75°   **E.** 413°   **F.** $\frac{264}{7}$ radians/sec   **G.** **(a)** 20π in.   **(b)** $50\pi - 100$ in²

Section 6.4

A. $1/\sqrt{2}$ **B.** $2/\sqrt{3}$ **C.** $-2/\sqrt{3}$ **D.** 0.6716

Section 6.5

A.

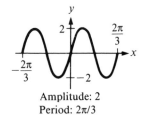

Amplitude: 2
Period: $2\pi/3$

B.

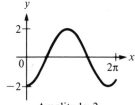

Amplitude: 2
Period: 6

C.

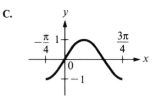

Amplitude: 1
Period: π
Phase shift: $-\pi/4$

D. $y = -4 \cos 10x$ **E.** **(a)** $3\pi/2$ **F.** **(a)** $\dfrac{1 - (-1)}{2} = 1$ **(b)** $\dfrac{|a| - (-|a|)}{2} = |a|$ **(c)** $\dfrac{3 - 1}{2} = 1$

Section 6.6

A.

x	0	$\pi/6$	$\pi/3$	$\pi/2$	$2\pi/3$	$5\pi/6$	π	$7\pi/6$	$4\pi/3$	$3\pi/2$	$5\pi/3$	$11\pi/6$	2π
y	und.	2	1.2	1	1.2	2	und.	−2	−1.2	−1	−1.2	−2	und.

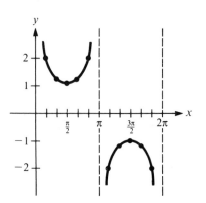

B. D: Set of all real numbers except $x = k\pi$ (k is any integer); R: $(-\infty,2] \cup [2,\infty)$; Period: 2π **C.** Increasing in all intervals

D.

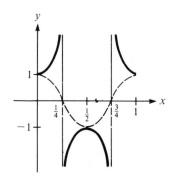

E.

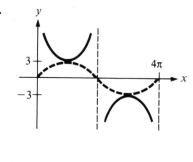

Section 6.7

A. $\cos x \sec x - \cos^2 x = 1 - \cos^2 x = \sin^2 x$ **B.** $\dfrac{1 + \dfrac{\cos^2 x}{\sin^2 x}}{\dfrac{\cos^2 x}{\sin^2 x}} = \dfrac{\sin^2 x + \cos^2 x}{\cos^2 x} = \dfrac{1}{\cos^2 x} = \sec^2 x$

C. $(\tan^2 x - \sec^2 x)(\tan^2 x + \sec^2 x) = (\tan^2 x - \tan^2 x - 1)(\tan^2 x + \sec^2 x) = -1(\tan^2 x + \sec^2 x)$

Section 6.8

A. $\cos \frac{1}{2}x \cos(\pi/2) + \sin \frac{1}{2}x \sin(\pi/2) = 0 + (\sin \frac{1}{2}x)(1) = \sin \frac{1}{2}x$ **B.** (a) $\frac{5}{13}$ (b) $-\frac{4}{5}$ (c) $\frac{33}{65}$ (d) $\frac{16}{65}$

C. $\begin{aligned}\sin 3x &= \sin(2x + x) = \sin 2x \cos x + \cos 2x \sin x\\ &= 2 \sin x \cos x \cdot \cos x + (\cos^2 x - \sin^2 x)(\sin x)\\ &= 2 \sin x \cos^2 x + \cos^2 x \sin x - \sin^3 x\\ &= 3 \sin x \cos^2 x - \sin^3 x\\ &= 3 \sin x (1 - \sin^2 x) - \sin^3 x\\ &= 3 \sin x - 3 \sin^3 x - \sin^3 x\\ &= 3 \sin x - 4 \sin^3 x\end{aligned}$

D. $\dfrac{1 - (1 - 2\sin^2 x)}{1 + (1 - 2\sin^2 x)} = \dfrac{2\sin^2 x}{2 - 2\sin^2 x} = \dfrac{\sin^2 x}{1 - \sin^2 x} = \dfrac{\sin^2 x}{\cos^2 x} = \tan^2 x$

Section 6.9

A. $\{5\pi/6, 7\pi/6\}$ **B.** $\{x: x = 0.32 + k2\pi \text{ or } x = 3.46 + k2\pi \ (k \text{ any integer})\}$ **C.** $\{\pi/12, 5\pi/12, 13\pi/12, 17\pi/12\}$ **D.** $\{0.83, 2.31, 3.71, 5.71\}$
E. $\{0, \pi\}$

Section 6.10

A. $-\pi/6$ **B.** $\sqrt{2}/2$ **C.** $\sqrt{2}/3$ **D.** $\arccos(x + 1) - (x + 1)\sqrt{1 - (x + 1)^2}$ **E.** $t = (\arccos a)/\pi$

Chapter 7

Section 7.1

A. $A = 53°, b = 6.9$ ft, $a = 45$ ft **B.** Data is inconsistent. **C.** 119 ft

Section 7.2

A. $A = 14°30', B = 38°50', C = 126°40'$ **B.** $C = 43°10', B = 76°20', a = 89.5$ ft **C.** 18.8 mi; 26.2 mi

Section 7.3

A. (a) $40\overline{0}$ ft/sec (b) 53.1° **B.** 3.8 lb (horizontal), 3.2 lb (vertical) **C.** 176 lb, 31° **D.** 420 lb

Section 7.4

A. $2(\cos 45° + i \sin 45°)$ **B.** $2 + 2i$ **C.** (a) $3(\cos 225° + i \sin 225°)$ (b) $3(\cos 45° + i \sin 45°)$
(c) $\frac{1}{3}(\cos(-45°) + i \sin(-45°))$ **D.** $-512i$ **E.** $\{(\sqrt{3}/2) + \frac{1}{2}i, (-\sqrt{3}/2) + \frac{1}{2}i, 0 - i\}$

Section 7.5

A.

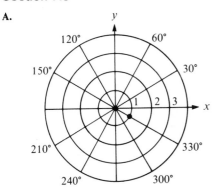

B. $(2, 495°); (2, -225°); (-2, -45°)$

C.

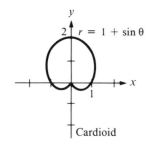

$r = 1 + \sin\theta$

Cardioid

D.

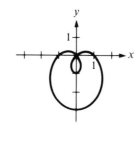

E. $(\sqrt{2}, \sqrt{2})$ **F.** $(6, 240°)$

G. $r = \cos\theta/\sin^2\theta$ or $r = \cot\theta\csc\theta$ **H.** $y = 1$

Chapter 8

Section 8.1

A. $16x + 2y + 13 = 0$ **B.** $-60x^2 + 4y^2 + 15 = 0$ **C.** $(3,2)$ **D.** $y = -\frac{3}{5}x + \frac{6}{5}$

Section 8.2

A. $(x + 1)^2 + (y - 5)^2 = 4$ **B.** $(-7,0); \sqrt{17}$ **C.** $(\frac{1}{2},1); \sqrt{13}/2$

Section 8.3

A.

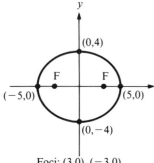

Foci: $(3,0), (-3,0)$

B.

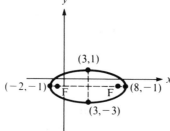

Foci: $(3 - \sqrt{21}, -1), (3 + \sqrt{21}, -1)$
Center: $(3, -1)$

C. $(x - 2)^2/25 + (y - 3)^2/9 = 1$ **D.** 32 ft

Section 8.4

A.

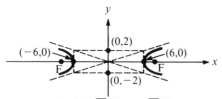

Foci: $(\sqrt{40},0), (-\sqrt{40},0)$
Asymptotes: $y = \pm\frac{1}{3}x$

B.

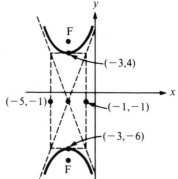

Foci: $(-3, -1 + \sqrt{29}), (-3, -1 - \sqrt{29})$
Center: $(-3, -1)$
Asymptotes: $y + 1 = \pm\frac{5}{2}(x + 3)$

C. $(x + 1)^2/9 - (y - 6)^2/72 = 1$

Section 8.5

A.

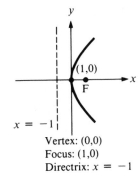

Vertex: (0,0)
Focus: (1,0)
Directrix: $x = -1$

B.

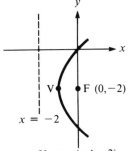

Vertex: $(-1,-2)$
Focus: $(0,-2)$
Directrix: $x = -2$

C. $(x - 5)^2 = -8(y - 1)$

D. **(a)** $x^2 = 12y$ if the vertex is at (0,0)   **(b)** 3 units from the vertex

Section 8.6

A. Hyperbola   **B.** Circle

Chapter 9

Section 9.1

A. $(-2,10)$ **B.** $(4,-1)$ **C.** Black, 12 g; Red, 20 g; $b + 6r = 132, 5b + r = 120$ **D.** **(a)** 4,000, B; $C = 3x + 3,000, C = 2x + 7,000$
(b) 3,000; $C = 4x, C = 3x + 3,000$

Section 9.2

A. 23   **B.** 0   **C.** (19,34)   **D.** $x = -3, y = 0, z = 2$

Section 9.3

A. $a = -3, b = -1;$ $\begin{bmatrix} 3 & -7 & | & -2 \\ 5 & 2 & | & -17 \end{bmatrix}$ **B.** $x = \frac{8}{3}, y = -2, z = \frac{1}{3}$ **C.** $x = 2, y = 3, z = 4, w = 5$
D. $x^2 + y^2 - 6x - 4y - 3 = 0$; center (3,2), radius 5

Section 9.4

A. 4×1 ("read 4 by 1")   **B.** $a = -1, b = 2, c = 4, d = \frac{1}{3}, e = 1, f = 2$ **C.** $\begin{bmatrix} -8 \\ \frac{7}{4} \\ 1 \end{bmatrix}$ **D.** AB is $4 \times 2, BA$ is undefined.

E. $\begin{bmatrix} 11 \\ 3 \end{bmatrix}$ **F.** $\begin{bmatrix} 19 & 0 & -10 \\ -9 & -8 & 13 \end{bmatrix}$ **G.** $2x - y + 3z = 0, 5y - 9z = 2, -4x + 3y - z = -1$ **H.** $\begin{bmatrix} -5 & 4 & -3 \\ 10 & -7 & 6 \\ 8 & -6 & 5 \end{bmatrix}$

I. $\begin{bmatrix} 1 & 2 & 3 \\ 2 & 3 & 4 \\ 3 & 4 & 6 \end{bmatrix} \begin{bmatrix} x \\ y \\ z \end{bmatrix} = \begin{bmatrix} 0 \\ 2 \\ 3 \end{bmatrix}; x = 3, y = 0, z = -1$

Section 9.5

A. $A = 2, B = -2, C = 1, D = -2$ **B.** $1 + \dfrac{3}{2x + 1} + \dfrac{-4}{(2x + 1)^2}$ **C.** $\dfrac{5}{x + 1} + \dfrac{-5x}{x^2 + x + 1}$

Section 9.6

A. $(-1,9), (1,-9)$ **B.** $(2,3), (2,-3), (-2,3), (-2,-3)$ **C.** $(-2,-3)$ **D.** 11 in., 5.5 in.
E. Alan: 6 hr, Bill: 12 hr, Cora: 3 hr; 1 hr 43 min

Section 9.7

A.

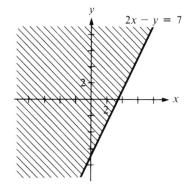

2x − y = 7

B. (2,5)

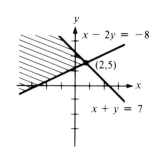

x − 2y = −8

(2,5)

x + y = 7

C. Max: 28, min: 16

D. $P = 78$ at $x = 6, y = 4$ **E.** $4x + 2y \geq 160; x + 2y \geq 100$ **F.** A: 20 ounces, B: 40 ounces, $C = 0.20x + 0.15y$

Chapter 10

Section 10.1

A. $2, 1, \frac{8}{9}, 1; \frac{1,024}{100}$ **B.** $AP; d = 1.01; 16.06, 17.07$ **C.** $a_n = (5n - 13)/9; \frac{37}{9}$ **D.** $a_n = (\frac{1}{4})^{n-1}; \frac{1}{16,384}$ **E.** $\frac{265}{12}$

Section 10.2

A. 87   **B.** $10^3 + 10^4 + 10^5 + 10^6 + 10^7; 11,111,000$ **C.** $\displaystyle\sum_{i=1}^{6} \frac{2i + 1}{3i + 1}$ **D.** 18 yrs   **E.** 26.394 ft

Section 10.3

A. $\frac{5}{28}$ **B.** $\frac{5}{33}$ **C.** $\{x: |x| < 1\}; S = \dfrac{2}{1 - x}$

Section 10.4

A. Step 1: $1 \cdot 2 \overset{?}{=} \dfrac{1(1 + 1)(1 + 2)}{3}; 2 = 2$

Step 2: $(1 \cdot 2) + (2 \cdot 3) + (3 \cdot 4) + \cdots + \underbrace{k(k + 1)}_{\text{induction assumption}} + (k + 1)[(k + 1) + 1]$

$= \dfrac{k(k + 1)(k + 2)}{3} + (k + 1)(k + 2)$

$= \dfrac{k(k + 1)(k + 2) + 3(k + 1)(k + 2)}{3}$

$= \dfrac{(k + 1)(k + 2)(k + 3)}{3}$

$= \dfrac{(k + 1)[(k + 1) + 1][(k + 1) + 2)]}{3}$

B. Step 1: If $n = 1$, $1^3 + 14(1) + 3 = 18$, which is divisible by 3.

Step 2: $(k + 1)^3 + 14(k + 1) + 3 = (k^3 + 3k^2 + 3k + 1) + (14k + 14) + 3 = (k^3 + 14k + 3) + (3k^2 + 3k + 15)$. Then $3k^2 + 3k + 15$ is obviously divisible by 3, while $k^3 + 14k + 3$ is divisible by 3 by the induction assumption. Thus, $(k + 1)^3 + 14(k + 1) + 3$ is divisible by 3.

Section 10.5

A. $81c^4 + 108c^3d + 54c^2d^2 + 12cd^3 + d^4$ **B.** $x^{10} + 20x^9 + 180x^8 + 960x^7$ **C.** $\binom{3}{0}, \binom{3}{1}, \binom{3}{2}, \binom{3}{3}$ **D.** 84

Section 10.6

A. 455   **B.** 15   **C.** 120   **D.** 5,005   **E.** 720   **F.** 128   **G.** $n = 9$

Section 10.7

A. (a) 0.40   (b) 0.60   (c) 0.40   (d) 0.70   **B.** $\frac{1}{216}$ **C.** $\frac{1}{60}$ **D.** (a) 0.01   (b) 0.81   (c) 0.18

Answers to Sample Test Questions

Chapter 1

1. **a.** $-\frac{21}{32}$ **b.** $3\sqrt{10}/20$ **c.** $(3\pi + \sqrt{5})/5$ **d.** -13 **e.** $\frac{3}{4}$ **f.** $\frac{49}{3}$ **g.** $2\pi\sqrt{14}$ **h.** $-\sqrt{2}$ **2. a.** $(-1 + 2\sqrt{2})/2$
b. 15 **c.** 5 **d.** -24 **e.** -1 **3. a.** 5 **b.** 5 **c.** $-\frac{1}{5}$ **4. a.** $7 - 4x$ **b.** $2a^3b^3 - 3a^2b^4 + 4a^2b^3$ **c.** $(1 - a)/(1 + a)$
d. a^x **e.** $x/(y^2z)$ **f.** $(a - b)^{1/4}$ **g.** $18(n + 1)$ **h.** $a + 3a^{1/3} + 3a^{-1/3} + a^{-1}$ **i.** $y^{1/8}$ **j.** $3x^3/(3x + a)$
k. $-2x^3 - x^2 + 9x + 9$ **l.** $3x^2 + 3xh + h^2$ **m.** $-(x + y)/xy$ **n.** $\dfrac{n + 3}{n}$ **o.** $\dfrac{6x^2 - 5}{15x}$ **p.** $\dfrac{6 - t}{3 - 2t}$ **q.** $\dfrac{y + x}{x}$
r. $x^3 + y^3$ **5. a.** $(x - 5)(x - 4)$ **b.** $3ax(3ax + 1)$ **c.** $n(n + 1)(n - 1)$ **d.** $3(k - 4)(k + 2)$
e. $(2x - y)(4x^2 + 2xy + y^2)$ **f.** $(x - 1)(x - 2)(x + 2)$ **6. a.** False **b.** False **c.** True **d.** False
7. **a.** $x^3 + 4x^2 + x - 6$ **b.** $x + 2\sqrt{xy} + y$ **c.** -1 **d.** $(b + a)/ab$ **e.** $\dfrac{5 - \sqrt{10}}{5}$ **f.** $1/\sqrt{x - 3}$ **8. a.** (b) **b.** (d)
c. (c) **d.** (a) **e.** (c) **f.** (e) **g.** (e) **h.** (e) **i.** (d)

Chapter 2

1. **a.** $\{\pm\frac{15}{2}\}$ **b.** $\{\frac{1}{2}\}$ **c.** $x = c/(a + b)$ **d.** $\{\pm i\}$ **e.** $\{24\}$ **f.** $\{(-5 \pm \sqrt{41})/4\}$ **g.** $\{3\}$ **h.** $\{-9\}$ **i.** $\{(2 \pm \sqrt{7})/(-3)\}$
2. **a.** $(\frac{2}{3},2)$ **b.** $(-3,4)$ **c.** $(-\infty,-7] \cup (-3,1)$ **d.** $(-\infty,\frac{3}{4})$ **3. a.** $4 + 6i$ **b.** $11 + 16i$ **c.** $(-4/5) - (8/5)i$
4. **a.** $-2\sqrt{2} + 0 \cdot i$ **b.** $-16 + 0 \cdot i$ **5. a.** 60 mph **b.** 5 sec **c.** 4.8 hours **d.** 1.62 sec **6. a.** $-2 - 3i$ **b.** $5i$ **c.** 1
d. 16 **e.** 25 **7. a.** (c) **b.** (b) **c.** (a) **d.** (a)

Chapter 3

1. **a.** $(-\infty,-1) \cup (1,\infty)$ **b.** $[-3,\infty)$ **c.** $-\frac{2}{3}$ **d.** $r = C/(2\pi)$ **e.** $1/x$ **f.** -2 **g.** $(0,1)$ **h.** $2\sqrt{29}$ **i.** $2x - 2$ **j.** $\frac{7}{2}$
2. **a.** 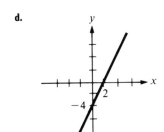 **b.** 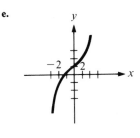 **c.**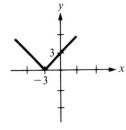

d. 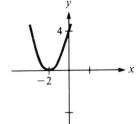 **e.** 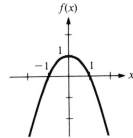 **f.**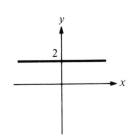

3. **a.** $(-\infty,\infty)$ **b.** $[-1,1]$ **c.** 0 **d.** $\{0, \pm 1, \pm 2\}$ **e.** $(-1,0) \cup (1,2)$ **4. a.** 0 **b.** $[-1,\infty)$ **c.** $[-1,1]$ **5.** $-4x - 2h$
6. $\frac{17}{2}$ **7. a.** 8 **b.** $t = 690 + 0.21(i - 4,000); [4,000,6,000]$ **8. a.** (d) **b.** (b) **c.** (c) **d.** (b) **e.** (e) **f.** (c) **g.** (a)
h. (e) **i.** (c)

Chapter 4

1. a. $(-\infty, 10]$ **b.** 6 **c.** 54 **d.** $2 - \sqrt{7}$ **e.** $x = 1$ **f.** $y = -1$ **g.** 0 **h.** $y = x(x - 2)(x + 3)$ **i.** $((1 \pm \sqrt{33})/2, 0)$
j. $\pm 2, \pm 1, \pm\frac{1}{3}, \pm\frac{2}{3}$ **k.** $-\frac{5}{4}$ **l.** $-\frac{1}{3}$ **2.** $f(x) = -\frac{7}{6}x + \frac{17}{6}$ **3. a.** $\pm\sqrt{6}/2$ **b.** ± 1 **c.** $3, 2, -1$ **4. a.** $[-3, 4]$
b. $(-\infty, \infty)$ **c.** $(-\infty, -\sqrt{2}) \cup (\sqrt{2}, \infty)$ **5. a.** Quotient: $x^2 + 2$, remainder: $-6x + 5$ **b.** Quotient: $3x^3 - 5x^2 + 5x - 4$, remainder:
-3 **6.** 277 **7.** $\frac{2}{3}$ and $-\frac{1}{2}$ (from solving $6x^2 - x - 2 = 0$) **8.** 3.1 **9. a.** $x = 0, x = -7$, hole at $x = 1$ **b.** $y = 0$,
crosses at $(\frac{3}{5}, 0)$ **10. a.** 400 (use $P = xy$ where $(x + y)/2 = 20$) **b.** $\{t: 1 \text{ sec} < t < 5 \text{ sec}\}$ **11. a.**

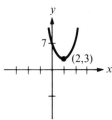

b.

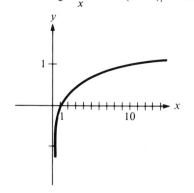

Vertical asymptote: $x = -1$
Horizontal asymptote: $y = 0$

c.

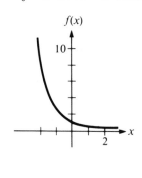

$h(x) = x + 1, x \neq 0$

12. a. (b) **b.** (b) **c.** (d) **d.** (b) **e.** (a)

Chapter 5

1. a. 3 **b.** 1 **c.** -1 **d.** 0 **e.** 2.0899 **f.** 4.8122 **g.** 0.4966 **h.** 25 **i.** 1.2263 **2. a.** $\log_8 2 = \frac{1}{3}$
b. $a^c = b$ **c.** $\log \dfrac{x + h}{x}$ **d.** $(a + b)/2$ **e.** 32 **f.** $\frac{1}{8}$ **g.** $(3, \infty)$ **h.** $(0, \infty)$ **i.** 0 **j.** $y = e^x$

3. a.

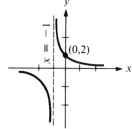

b.

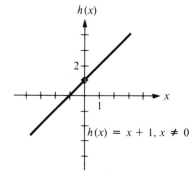

4. $(2, \frac{1}{81}), (0, 1), (-1, 9), (-2, 81), (\frac{3}{2}, \frac{1}{27})$ **5. a.** $\{4\}$ **b.** $\{2.5480\}$ **c.** $\{100\}$ **d.** $\{5.7833\}$ **e.** $\{4.0146\}$ **f.** $\{6\}$ **g.** $\{-19.9279\}$
h. $\{5\}$ **i.** $\{3\}$ **6. i.** a, b, d **ii.** b **iii.** e **7. a.** \$4,407.98 **b.** 10.99 percent **c.** 4.3 days **d.** 1,000 **8. a.** (e) **b.** (b)
c. (c) **d.** (b)

Chapter 6

1. a. $\sqrt{3}/2$ **b.** -11.47 **c.** -1.590 **d.** $1/\sqrt{2}$ **3.** 0 **f.** 1.556 **g.** 0 **h.** 0.1395 (Calc: 0.1411) **i.** 1.557 **j.** $\frac{1}{2}$
k. 0.32 **l.** Undefined **2. a.** $-\frac{1}{3}$ **b.** $\frac{3}{2}$ **c.** $-\sqrt{3}/2$ **d.** 2 and 3 **e.** 36 **f.** 370° (other possibilities) **g.** 67° **h.** 80°
i. Set of all real numbers except $x = (\pi/2) + k\pi$ (k is any integer) **j.** $[-\pi/2, \pi/2]$ **k.** $2\pi/3$ **l.** $2\pi/5$ **m.** $11\pi/6$ **n.** 135°
o. $\pi/8$ **p.** 12π in.² **q.** $1 + \arcsin a$ **3. a.** $\{51°\}$ **b.** $\{315°\}$ **c.** $\{\theta : \theta = k360°, k \text{ is any integer}\}$ **d.** $\{26°\}$ **e.** $\{25°, 155°\}$
f. $\{93°\}$ **4. a.** $\{x : x = 1.85 + k2\pi \text{ or } x = 4.99 + k2\pi, k \text{ is any integer}\}$ **b.** $\{\pi/3, 2\pi/3\}$ **c.** $\{0, \pi/2, \pi\}$ **d.** $\{\pi/4, 7\pi/4\}$ **5.** 1,040 ft

6. a. $\dfrac{\cos^2 x}{\sin^2 x} + 1 = \dfrac{\cos^2 x + \sin^2 x}{\sin^2 x} = \dfrac{1}{\sin^2 x} = \csc^2 x$

   b. $\dfrac{1 - \cos 2x}{2} = \dfrac{1 - (1 - 2\sin^2 x)}{2} = \dfrac{2\sin^2 x}{2} = \sin^2 x$

   c. $\cos\dfrac{3\pi}{2} \cos x - \sin\dfrac{3\pi}{2} \sin x = (0)\cos x - (-1)\sin x = \sin x$

7. $\arcsin x + 2x\sqrt{1 - x^2}$ **8. a.** $1/a$ **b.** $1 - 2a^2$ **c.** x **9. a.**

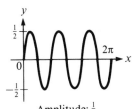

Amplitude: $\frac{1}{2}$
Period: $2\pi/3$

b.

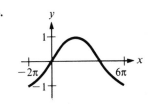

Amplitude: 1
Period: 8π
Phase shift: -2π

c.

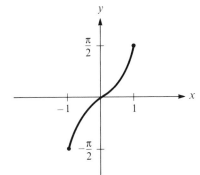

10. a. (a) **b.** (b) **c.** (d) **d.** (a) **e.** (b) **f.** (d) **g.** (e) **h.** (a) **i.** (b)

Chapter 7

1. 6.4 m **2.** 13° **3.** 5.5 lb **4. a.** $\sqrt{2}(\cos 315° + i \sin 315°)$ **b.** $2(\cos 225° + i \sin 225°)$ **c.** $\cos 90° + i \sin 90°$ **5.** $(6, \pi/6)$
6. $(-\sqrt{3}, 1)$ **7.** $r = 1 - \cos\theta$ **8. a.**

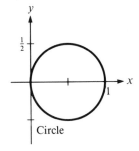

Circle

b.

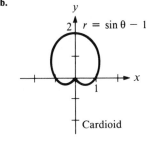

$r = \sin\theta - 1$

Cardioid

9. a. (c) **b.** (d) **c.** (d) **d.** (a) **e.** (b) **f.** (a) **g.** (c)

Chapter 8

1. a. $(x + 2)^2 + (y - 4)^2 = 9$ **b.** $x^2 = 32y$ **c.** Ellipse   **d.** $y = \pm\frac{1}{2}x$ **e.** $\sqrt{23}$ **f.** $(-5,5)$ **g.** Parabola
h. $(x^2/33) + (y^2/49) = 1$ **i.** $(\pm\sqrt{29},0)$ **j.** Circle   **2. a.**

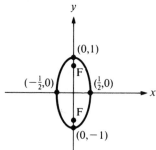

Foci: $(0,\sqrt{3}/2), (0,-\sqrt{3}/2)$

b.

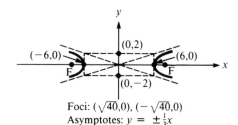

Foci: $(\sqrt{40},0), (-\sqrt{40},0)$
Asymptotes: $y = \pm\frac{1}{3}x$

c.

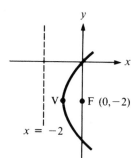

Vertex: $(-1,-2)$
Focus: $(0,-2)$
Directrix: $x = -2$

3. $0 = x - y + 1$ **4. a.** 23.6 ft   **b.** 22.2 ft   **5. a.** (a)   **b.** (a)   **c.** (c)   **d.** (b)   **e.** (d)

Chapter 9

1. a. 10   **b.** -46 **2. a.** Undefined   **b.** $\begin{bmatrix} -3 & 6 \\ 1 & -5 \end{bmatrix}$ **c.** $\begin{bmatrix} 2 & 6 & 8 \\ -2 & 0 & 12 \end{bmatrix}$ **d.** $\begin{bmatrix} -1 & -3 & -4 \\ -4 & 0 & -24 \\ 4 & 6 & 20 \end{bmatrix}$ **3.** $\begin{bmatrix} -\frac{4}{11} & \frac{3}{11} \\ \frac{1}{11} & \frac{2}{11} \end{bmatrix}$
4. a. $x = 1, y = 1$ **b.** $x = \frac{1}{2}, y = \frac{2}{3}, z = -\frac{5}{6}$ **5. a.** $(3,1), (3,-1), (-3,1), (-3,-1)$ **b.** $(-2,4)$ **c.** $x = 2, y = 3, z = -4$
6. a. $-2,4$ **b.** 311   **7.**

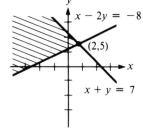

$x - 2y = -8$
$(2,5)$
$x + y = 7$

8. $A = 1, B = 2, C = -2$ **9. a.** $\dfrac{4}{2x + 3} + \dfrac{-1}{x - 1}$

b. $\dfrac{-1/3}{x} + \dfrac{(7/3)x}{x^2 + 3x + 3}$ **10.** Max $P = 5$ at $x_1 = 0, x_2 = 5$ **11.** 8 lb of A, 3 lb of B; \$2.80 per day

Chapter 10

1. a. $-75x^{14}y$　**b.** $a_n = 0.3n + 0.8$　**c.** 1, 1, 2, 3, 5　**d.** -1 and 1　**e.** 120　**f.** $3^2 + 3^3 + 3^4$　**g.** $\frac{1}{10}$　**h.** Geometric progression

i. $\frac{1}{36}$　**j.** $\displaystyle\sum_{i=0}^{3} x^i$　**2.** $16x^4 - 96x^3y + 216x^2y^2 - 216xy^3 + 81y^4$　**3. a.** $(1.06^{10} - 1)/0.06 = 13.18$　**b.** 12.7　**c.** 10　**d.** 12

4. 5,050　**5. a.** 5　**b.** 3　**6.** 120　**7.** 56　**8.** 480　**9.** 24

10. Step 1: $\dfrac{1}{1\cdot 3} \overset{?}{=} \dfrac{1}{2(1)+1}$; $\dfrac{1}{3} = \dfrac{1}{3}$.

Step 2: Assume $\dfrac{1}{1\cdot 3} + \dfrac{1}{3\cdot 5} + \dfrac{1}{5\cdot 7} + \cdots + \dfrac{1}{(2k-1)(2k+1)} = \dfrac{k}{2k+1}$.

Then $\dfrac{1}{1\cdot 3} + \dfrac{1}{3\cdot 5} + \dfrac{1}{5\cdot 7} + \cdots + \dfrac{1}{(2k-1)(2k+1)} + \dfrac{1}{[2(k+1)-1][2(k+1)+1]}$

$= \dfrac{k}{2k+1} + \dfrac{1}{[2(k+1)-1][2(k+1)+1]} = \dfrac{k(2k+3)+1}{(2k+1)(2k+3)}$

$= \dfrac{(k+1)(2k+1)}{(2k+1)(2k+3)} = \dfrac{k+1}{2k+3} = \dfrac{k+1}{2(k+1)+1}$.

11. a. 0.04　**b.** 0.64　**c.** 0.32　**12. a.** (c)　**b.** (b)　**c.** (a)　**d.** (e)　**e.** (e)　**f.** (b)　**g.** (a)　**h.** (d)　**i.** (c)

Trigonometry

General Angle Definitions

If θ is an angle in standard position and if (x,y) is any point on the terminal ray of θ [except $(0,0)$], then

$$\sin \theta = \frac{y}{r} \quad \leftarrow \text{reciprocals} \rightarrow \quad \csc \theta = \frac{r}{y}$$

$$\cos \theta = \frac{x}{r} \quad \leftarrow \text{reciprocals} \rightarrow \quad \sec \theta = \frac{r}{x}$$

$$\tan \theta = \frac{y}{x} \quad \leftarrow \text{reciprocals} \rightarrow \quad \cot \theta = \frac{x}{y}.$$

Note: $r = \sqrt{x^2 + y^2}$

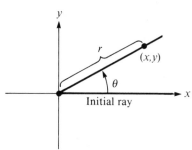

Unit Circle Definitions

Consider a point (x,y) on the unit circle $x^2 + y^2 = 1$ at arc length s from $(1,0)$. Then

$$\sin s = y \quad \leftarrow \text{reciprocals} \rightarrow \quad \csc s = \frac{1}{y}$$

$$\cos s = x \quad \leftarrow \text{reciprocals} \rightarrow \quad \sec s = \frac{1}{x}$$

$$\tan s = \frac{y}{x} \quad \leftarrow \text{reciprocals} \rightarrow \quad \cot s = \frac{x}{y}.$$

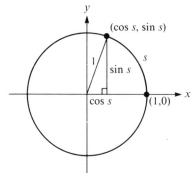

Right Triangle Definitions

If θ is an acute angle in a right triangle as shown below, then

$$\sin \theta = \frac{\text{opposite}}{\text{hypotenuse}} \quad \leftarrow \text{reciprocals} \rightarrow \quad \csc \theta = \frac{\text{hypotenuse}}{\text{opposite}}$$

$$\cos \theta = \frac{\text{adjacent}}{\text{hypotenuse}} \quad \leftarrow \text{reciprocals} \rightarrow \quad \sec \theta = \frac{\text{hypotenuse}}{\text{adjacent}}$$

$$\tan \theta = \frac{\text{opposite}}{\text{adjacent}} \quad \leftarrow \text{reciprocals} \rightarrow \quad \cot \theta = \frac{\text{adjacent}}{\text{opposite}}.$$

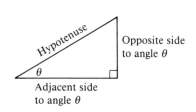

Signs of the Trigonometric Ratios

$$\left.\begin{array}{l}\sin \theta \\ \csc \theta\end{array}\right\} \text{positive} \qquad \underline{A}\text{ll the functions are positive}$$

others} negative

$$\left.\begin{array}{l}\tan \theta \\ \cot \theta\end{array}\right\} \text{positive} \qquad \left.\begin{array}{l}\cos \theta \\ \sec \theta\end{array}\right\} \text{positive}$$

others} negative $\qquad$ others} negative